About Island Press

Island Press is the only nonprofit organization in the United States whose principal purpose is the publication of books on environmental issues and natural resource management. We provide solutions-oriented information to professionals, public officials, business and community leaders, and concerned citizens who are shaping responses to environmental problems.

In 1994, Island Press celebrated its tenth anniversary as the leading provider of timely and practical books that take a multidisciplinary approach to critical environmental concerns. Our growing list of titles reflects our commitment to bringing the best of an expanding body of literature to the environmental community throughout North America and the world.

Support for Island Press is provided by The Geraldine R. Dodge Foundation, The Energy Foundation, The Ford Foundation, The George Gund Foundation, William and Flora Hewlett Foundation, The John D. and Catherine T. MacArthur Foundation, The Andrew W. Mellon Foundation, The Joyce Mertz-Gilmore Foundation, The New-Land Foundation, The Pew Charitable Trusts, The Rockefeller Brothers Fund, The Tides Foundation, Turner Foundation, Inc., The Rockefeller Philanthropic Collaborative, Inc., and individual donors.

FUTURE DRIVE

FUTURE DRIVE

Electric Vehicles and
Sustainable Transportation

Daniel Sperling

WITH CONTRIBUTIONS FROM

MARK A. DELUCCHI,

PATRICIA M. DAVIS, AND

A.F. BURKE

ISLAND PRESS
Washington, D.C. • Covelo, California

ISLAND PRESS is a trademark of The Center for Resource Economics.

Library of Congress Cataloging-in-Publication Data

Sperling, Daniel.
    Future drive : electric vehicles and sustainable transportation /
Daniel Sperling ; with contributions from Mark A. Delucchi, Patricia
M. Davis, and A.F. Burke.
        p.   cm.
    Includes bibliographical references (p.   ) and index.
    ISBN 1-55963-327-1 (alk. paper). — ISBN 1-55963-328-X (pbk. :
alk. paper)
        1. Automobiles, Electric.   2. Transportion, Automotive—
Environmental aspects.   3. Transportation and state.   I. Delucchi,
Mark A.   II. Davis, Patricia M.   III. Burke, A.F. (Andrew F.)
IV. Title.
TL220.S65   1995
388—dc20   94-38935
                                                                CIP

Printed on recycled, acid-free paper  ∞

Manufactured in the United States of America

10   9   8   7   6   5   4   3   2   1

To Rhiannon and all the children of her generation

Contents

Preface

It was with some apprehension that I began this book. None of my previous books and papers strongly promoted any particular position, and therefore none of them seriously threatened business interests, government missions, or environmental lobbies. Even after fifteen years of research into a variety of transportation, energy, and environmental topics, no particular transportation or energy strategy stood out in my mind as meriting strong and immediate action. Recently, however, it has become clear to me that the small incremental improvements in the environmental and economic performance of transportation systems of past decades are being overwhelmed by the rapid growth in people, cars, and travel. To remain satisfied with incremental improvements is to accept a slow deterioration in environmental and urban conditions. It is time for society to contemplate a more radical break from the transportation and energy strategies of the past.[1]

Two events inspired this book and its more assertive stance. One was my participation in a two-year study of the "future of the auto," chaired by Elmer Johnson, a lawyer and former high-ranking executive of General Motors. Our research panel was bombarded with the economist's mantra, "Get the price right," the presumption being that all solutions would follow from there. But getting the price right means raising taxes and fees, and even the most naive political observer recognizes the lack of political support for further government regulation of the price of fuel and vehicles. (One recent example is President Clinton's proposal, in early 1993, to raise the gasoline tax 8 cents per gallon; a national uproar followed, ultimately resulting in a paltry 4.3-cent increase.)

What our panel needed was a broader, more realistic vision; Chairperson Johnson was the only one of our group who offered a clear one. He argued that transportation was a fundamental cause of, and potential solution to, the breakdown of community in America's inner cities and edge cities alike. He doubted that technology could solve all the social ills caused by the loss of community. Rather, he believed the key to recovery was less driving, and he framed his vision around that precept.[2] Johnson's idea was certainly appealing; less use of cars would certainly solve many problems. But our panel could not muster evidence to justify such a sweeping indictment of the automobile—a variety of other factors struck us as also responsible for social breakdown. Moreover, panel members pointed out, any proposal to curb driving dramatically would be politically naive, especially in affluent countries. Our final report reflected the disparate views of the panel more than the social vision of the chair.[3] This book is partly a response to Elmer Johnson and my economist colleagues; it offers what I believe to be a coherent and compelling alternative vision.

The second motivation was more personal: the birth of my daughter. Until then I had, like most policy analysts, treated events more than twenty years in the future as virtually irrelevant. Then suddenly I realized that not only would my daughter be alive in twenty years, but her life would have barely begun. Even more eye opening was the realization that in 2050, off the chart of most economists and policy analysts (and business people and politicians), she won't even have reached retirement age. Creating a livable world in 2050 was no longer irrelevant, or just an academic exercise. And therein lies whatever bias I may have in writing this book: that we are morally obligated to guard with diligence the interests of future generations by actively seeking to preserve the long-term environmental and economic viability of this planet.[4]

This book took about a year to write, but whatever knowledge and insights it might impart were formed over a much longer time. They are the result of years of research on topics ranging from consumer demands with respect to fuel and cars to the design of fuel cell vehicles. On several chapters I was fortunate to have the assistance of three talented people: Dr. Andrew Burke, now with me at the University of California, Davis, a mechanical and electrical engineer internationally recognized for his research on electric-vehicle technology; Patricia Davis, a former finance manager and economist in transportation and energy, who provided insights into business thinking and who, along with my daughter, alerted me to the importance of the long-term future; and Dr. Mark Delucchi, the quintessential interdisciplinarian, an expert in engineering, economics, and ecology, and internationally known for his detailed analyses of the costs and environmental impacts of motor vehicles and transportation fuels. The diverse expertise of these three individuals suggests the range of knowledge and experience needed to understand transportation choices and to untangle facts from values, ideology, and vested interests. This book could not have been written without them, or without the research conducted at the University of California, Davis, by a variety of graduate students and faculty researchers ranging from mechanical engineers to anthropologists.

But this work is not solely a product of academe. As director of the Institute of Transportation Studies at the University of California, Davis, I regularly interact with experts from industry, government, environmental groups, and academia. These acquaintances—including corporate supporters of the institute such as Chevron, Bank of America, Nissan, Exxon, and California electric utilities, a variety of federal and state (California) agencies, environmental groups such as the Environmental Defense Fund, Natural Resources Defense Council, and the Union of Concerned Scientists, and various legislative leaders in Washington, D.C., and Sacramento—have scrutinized the findings and proposals that follow.

What I have learned from this broad acquaintance is that no single set of rules can be devised to solve one of the most vexing issues of modern life—how to counteract the ill effects of automobile proliferation without curtailing freedom of movement and choice. What we need is a compelling vision of what the future should look like. Although I reject ideology—a rigid set of beliefs that require internal consistency at the expense of reality—I am a strong proponent of vision. We must be able to envision the destination before starting the trip, and meanwhile constantly monitor technological developments, market preferences, and the effect of government policies and regulations without letting such details bog us down. Without a central vision and guiding principles, policymakers and business people will be overwhelmed by the bewildering array of choices and combinations in the fuel and vehicle industries. They will be unable to respond to special interests or to formulate strategic business plans.

Not everyone agrees with me on the importance of vision, as this provocative headline in the *Wall Street Journal* of October 4, 1993 suggests: "Robert Eaton [CEO of Chrysler] Thinks Vision Is Overrated and He's Not Alone." Eaton justified his startling claim by pointing out that the car industry is "mature." If this were true, then it would be perfectly reasonable for Eaton to preoccupy himself with, in his words, "quantifiable short-term results." But the automotive industry is *not* mature; it is on the brink of technological revolution. Leaders such as Eaton are well aware of the bubbling cauldron of technological innovation in the transportation industry, and it makes them apprehensive: the fledgling technologies that hold the most promise for curbing energy consumption and air-polluting emissions also have the most radical implications for the automotive industry.

It will not be the invisible hand of the marketplace that brings these technologies into being. Opposition by America's Big Three automakers (and others) to California's recent zero-emission-vehicle mandate is a calculated strategy to hinder innovation. While corporate leaders may prefer to suppress new technologies, the combination of a strong societal commitment to an environmentally healthier world and the intense state of competition in the industry suggests that change will come—sooner rather than later. If business leaders need vision to guide their companies through the jungle of intensifying competition, government leaders need vision—and the means of communicating it to a vast number of political, administrative, business, and intellectual leaders—if they are to have any hope of effecting major change and solving the many problems associated with proliferating automobiles.

A vision becomes reality only if it is compelling to a wide range of interests. This book is an attempt to create a democratic vision. My aim is to suggest how we might redirect the transportation system toward environmental sustainability without forcing consumers to drive less.[5]

Acknowledgments

Many people have had a hand in this book. Foremost among them are the three "contributors": Andrew Burke, Patricia Davis, and Mark Delucchi. Andy and I fully collaborated on Chapter 6. It was a shared intellectual undertaking that gave me tremendous respect for his knowledge and insight. I look forward to many years of collaborative research with Andy now that he has joined us at the Institute of Transportation Studies at the University of California, Davis. Patricia Davis's contribution, as my marriage and intellectual partner, pervades the entire book. Her powerful logic in organizing ideas and her intolerance for sloppy thinking and technical jargon kept me alert. She stimulated my thinking with helpful suggestions and sensible objections. Mark Delucchi provided the initial inspiration for this book. In the late 1980s, while I was absorbed in research on liquid and gaseous alternative fuels, he encouraged me to focus more on electric vehicles. Together, we have. Much of his analytical work on costs and environmental impacts for battery and fuel cell vehicles is the foundation of this book. He made especially large contributions to Chapters 3 and 5. (Note that Mark Delucchi changed the spelling of his last name in 1993. In publications before that date, the name is spelled DeLuchi.)

To a large extent, this book is the product of research conducted by an outstanding group of graduate students and faculty researchers at ITS-Davis, including Kenneth Kurani, Tim Lipman, Marshall Miller, Jonathan Rubin (now at the University of Tennessee), Aram Stein, David Swan, Tom Turrentine, and Michael Quanlu Wang (now at Argonne National Laboratory), as well as Mark Delucchi. While their work is cited through the book, I would like to acknowledge here the special joy I take in our close working relationships and my pride in their achievements.

The manuscript was greatly enhanced by the large number of people (in addition to my UC Davis colleagues) who carefully reviewed one or more chapters for accuracy and content. They include Mary Brazell, Terry Day, William Falik, Deborah Gordon, John (Jay) Harris, Wendy James, Larry Johnson, Cece Martin, William McAdam, Michael Replogle, Sam Romano, Richard Schweinberg, Vito Stagliano, David Swan, Mel Webber, and Stein Weissenberger. I am grateful to them for sharing their time and expertise.

I am equally grateful to a number of distinguished researchers and thinkers who have stimulated me to go beyond the conventional wisdom.

They include Lon Bell, Tom Cackette, Elizabeth Deakin, William Garrison, Pat Grimes, Genevieve Guiliano, David Greene, Greig Harvey, Elmer Johnson, Ryuichi Kitamura, Charles Lave, Amory Lovins, Paul MacCready, Michael Replogle, Lee Schipper, and Martin Wachs. Each of them has been a continuing source of inspiration and a generous source of knowledge.

I am especially grateful to and admiring of Susie O'Bryant and Carol Earls for being so extraordinarily helpful and competent in running ITS-Davis during my extended absences (as well as when I was present). Along with Randy Guensler and Simon Washington, who so ably taught my classes for me, and my colleague Paul Jovanis, they made it possible for me to devote large amounts of time to the book.

I am also grateful to Resources for the Future for providing financial support (the Gilbert F. White Fellowship) and a stimulating intellectual refuge during the fall of 1993. Thanks especially to Robert Fri, Paul Portney, and Alan Krupnick for their enthusiasm and support, and to Chris Mendez for solving all those small problems that arise when away from one's own office.

I also thank Cary Sperling for her editing help, Glen Silber for suggesting the title of the book, Jon Vranesh for tracking down references, and Nancy Olsen and Heather Boyer, my editors, for their support and diligence in responding to a sometimes-distraught author, and to Constance Buchanan for superb copy editing.

Lastly, I thank the following organizations for financially supporting the research from which this book was created: the California and U.S. Departments of Transportation, California Air Resources Board, Chevron, Nissan, Exxon USA, Volvo, California Institute for Energy Efficiency, University of California Transportation Center, Energy Foundation, Pew Charitable Trusts, Pacific Gas & Electric, and Calstart. They, of course, do not necessarily endorse the findings of this book.

Two chapters of this book have been or will be published in different formats elsewhere. Chapter 4 is an abridged version of a paper, "Prospects for Neighborhood Electric Vehicles," forthcoming in *Transportation Research Record;* Chapter 6, "Hybrid Vehicles: Always Second Best?," is an abridged version of a report with the same title published in July 1994 by the Electric Power Research Institute.

CHAPTER 1

Transportation as if People Mattered

I am going to democratize the automobile. When I'm through everybody will be able to afford one, and about everyone will have one.

Henry Ford

We [human beings] are a big mistake. We're a case of gigantism. Our brains are too big and they're killing us. We've created all these poisons, which are unknown anywhere else in the universe. Of course, we want our brains to become even bigger so we can increase our supply of ideas. That's like elephants in trouble saying, "I think we'll be okay if we put on another hundred pounds."

Kurt Vonnegut

The world's car population is booming. Cars are polluting the world's cities, dumping increasing amounts of carbon dioxide and other climate-altering greenhouse gases into the atmosphere, and consuming vast quantities of petroleum. "The American dream of a car—or two or three of them—in every garage is beginning to look like a nightmare for our planet," warns the former president of the World Resources Institute.[1] Is this true? Will our thirst for automobile mobility inevitably lead to environmental and economic cataclysm? Many believe so, and with some justification.

The alarming reality is that the automobile population is growing at a much faster rate than the human population, with saturation nowhere in sight. In 1950, there were approximately 50 million vehicles on Earth, roughly 2 for every 100 persons. By 1994 the vehicle population had soared to almost 600 million, roughly 10 per 100 people. If present trends continue, over 3 billion vehicles could be in operation by the year 2050, exceeding 20 per 100. Even then, world car ownership rates would fall far short of current U.S. rates of 70 per 100 people.[2]

A sobering assessment of the future? Yes. Is disaster inevitable? Not necessarily. The future need not be a simple extrapolation of the past;

public policy can be changed and private investment altered. We can shift to a more environmentally friendly transportation system without unduly restricting freedom of movement. I envision a future only a few decades distant in which petroleum consumption, air pollution, and greenhouse gas emissions by new motor vehicles are reduced to near zero—at little or no additional cost.

Taming the Auto

What tools are available to craft this more benign future?[3] I start with certain premises—that it is important to respect people's preferred mode of travel; that we need greater diversity in vehicle and energy technologies; that government should take an active role in encouraging new technologies and industries (through more incentives and more flexible regulations); and that electric propulsion is central to this more benign future. Electric vehicles of various sizes and designs—including those powered by fuel cells—significantly reduce air pollution, greenhouse gases, and petroleum use. When powered by batteries they are especially suited to very small vehicles in a way that could expand mobility for many people and be a catalyst for reclaiming neighborhood streets for the use and enjoyment of people. Electric propulsion provides by far the best opportunity to create an environmentally benign transportation system.

Happily, we are not unrealistically far from realizing this vision. The stage was set in 1990 when the California Air Resources Board (CARB) galvanized a range of automotive and technology companies with what has become known as the zero-emission vehicle (ZEV) mandate. This radical new program requires that a specified percentage of manufacturers' sales consist of ZEVs. The ZEV mandate may be the single most important event in the history of transportation since Henry Ford began mass-producing cars eighty years ago. It set into motion a series of events that may revolutionize motor vehicles and perhaps transform the transportation system and motor vehicle manufacturing industry.

The California ZEV mandate has also been embraced by New York and Massachusetts and is being seriously considered by a number of other states. It is set to take effect in 1998 in California and shortly thereafter elsewhere.

The ZEV mandate promises to overshadow a trio of sweeping national laws passed in the early 1990s. These three laws—the Clean Air Act Amendments of 1990, the Intermodal Surface Transportation Efficiency Act (ISTEA) of 1991, and the Energy Policy Act of 1992—were conceived as the cornerstone of efforts to put a lid on pollution and, by redirecting the transportation system, to make possible a more environmentally benign future. They represent a complex of policies and programs designed to reduce solo driving, enhance local flexibility in dealing with transportation problems, introduce alternative fuels, and incorporate "intelligent" technology into the transportation system.

The progress of these laws has been spotty. Although they provide some direction, they have two serious flaws: ineffectiveness in reducing automobile use, and a failure to embrace electric propulsion.

A call for less dependence on cars is appropriate and desirable. Greater use of public transit, walking, bicycling, and telecommuting should be encouraged. But travel reduction will be difficult to accomplish and, beyond a point, is not even desirable. In any case, cars by themselves are not deterrents to a sustainable transportation and energy system. Worse, those advocating fewer cars are likely to be indifferent to the urgent need to build clean and efficient vehicles. This is certainly the sentiment expressed in a widely distributed report sponsored by the Aspen Institute and the American Academy of Arts and Science. Written by a former high-ranking executive of General Motors in collaboration with a blue ribbon committee, it argues that the American love of the automobile has "atomized urban life and stunted people's capacities to nurture and value shared forms of life: family, community and civic life."[4] This calls for "changes in policies and social habits that represent nothing less than a new transportation ethic for the twenty-first century."[5]

Advocates of this mind see automotive technology as demeaning to the human spirit and destructive of urban landscapes, turning them into sterile, lonely places studded with parking lots.[6] They resist solutions that would improve vehicles as implicit endorsements of a more central role for cars. To prevent expanded use of vehicles, extremists seem to imply that it would be preferable not to reduce vehicular pollution at all. Taming the car would embed it even more deeply in our lives. Such fear is out of touch with reality. There can be no turning back to a preautomotive age, nor is there any need to. With creativity and political will, the car can be made more benign, it can increase mobility, and it can even be a tool to help restore the human face of our communities.

How Much Does It Really Cost to Drive a Car?

The public debate is skewed toward the ways in which cars threaten the fabric of modern life. Yes, cars pollute, deplete petroleum, affect climate, and consume space. But cars also provide large benefits.

First, consider the mistaken belief that drivers pay only a small fraction of the true cost of driving. According to one report,

> Commuters going to work in major central business districts in the U.S. in their own motor vehicles directly pay for only about 25 percent of the total cost of their transport. The other 75 percent is typically borne by their employers (e.g., in providing "free" parking), by other users (in increased congestion, reduced safety, etc.), by fellow workers or residents (in air or noise "pollution," etc.) and by governments (passed on to the taxpayers of one generation or another in ways that usually bear no relationship to auto use).[7]

This claim may be true in a few especially polluted and congested downtowns, but it is not an accurate generalization.

The Congressional Office of Technology Assessment examined this question with the most detailed and rigorous analysis ever conducted of the social costs of motor vehicles. The conclusion? Motor vehicle users in the United States pay 68 to 80 percent of the total cost of motor vehicle use.[8] What they pay for are direct costs, including the car itself, fuel and fuel taxes, registration fees, insurance, and parking. What they don't pay for in full, among other things, is traffic congestion, accidents, the cost of buying oil from a cartel (measured as the effect on world oil prices), national security costs associated with oil importation, environmental degradation, and traffic courts. Only in certain situations, such as driving at peak hours in polluted and congested downtown New York, is car use heavily subsidized—that is, only at those times and places are drivers paying only a small share of the total cost.

In general, it would be desirable to price cars and their use to absorb unpaid costs. The result would be somewhat reduced driving, along with less congestion, pollution, and energy use. The effect might be significant in some regions—if it were politically possible to charge motorists higher fuel taxes, registration fees, parking charges, and so on. But given the benefits of the car, these unpaid costs are not high enough to justify a radical restructuring of transportation systems and lifestyles.

It is also not true that U.S. fuel prices are wildly distorted. Many believe that the price of gasoline should be much higher to account for oil spills, leaking storage tanks, pollution, global warming, energy dependency, and other unpaid extraction, transport, and combustion costs. Studies that have put the unpaid social cost of gasoline at over two dollars a gallon suffer from major methodological and analytical shortcomings.[9] For the Office of Technology Assessment's far more detailed and sophisticated social cost study, Mark Delucchi produced a more modest figure:

between 30 cents and $1.60 per gallon, and probably closer to the lower number.[10]

This is not to say that gasoline is an acceptable fuel. The problem is the tyranny of averages. Averages are misleading in this case because dramatic differences in pollution from one locale to another lead to large variations in environmental costs. One more ton of pollution in Los Angeles, for instance, will cause far more damage to health, buildings, trees, and crops than one more ton in South Dakota.

In California cities where pollution levels are high, emissions from each gallon of burned gasoline cause an average of about $1.50 worth of damage.[11] In less polluted cities such as Boston, equivalent emissions cause about 75 cents of damage. In rural areas, polloution damage is near zero. But even these cost estimates understate the problem. That is because each additional ton causes more damage than the average ton; this marginal effect is especially great in more polluted areas. The damage caused by an additional gallon of gasoline is considerably more than $1.50 in most California cities. Accordingly, it was economically rational that California factories, refineries, and other stationary pollution sources were spending about $2.50 per gallon in government-mandated programs in the early 1990s to eliminate the amount of pollution caused by each gallon of gasoline.[12]

In any case, it is clear that vehicles with low emissions should be highly valued in some regions. The challenge is to create public policies and rules that would introduce such vehicles and fuels in an efficient and effective manner.

So far, we have addressed *measurable* costs. What about other difficult-to-measure social costs often blamed on the auto—urban decay, loss of community, and marginalization of the poor, the elderly, and the disabled?[13]

It is certainly true that the rapid proliferation of automobiles has been a major influence on urban and suburban landscapes since the turn of the century. There is plenty of evidence that human settlements, not only in the United States but around the world, have become auto-centric, with urban design, infrastructure, and policymaking focused on accommodating the car. Today in many suburban communities, for example, sidewalks have been eliminated in the interests of providing more room for cars; this absence of sidewalks in turn encourages still greater dependence on cars by making walking more dangerous. Parents choose to drive their children to the park even if it is only two blocks away.

But blaming the automobile for this problem is like blaming the messenger for the bad news. The true blame lies with urban planning and governance—with people, not cars. It is not necessary to gouge out the urban landscape, to produce urban designs so hostile to pedestrians and cyclists, in order to accommodate cars.

A more fundamental criticism of the proliferation of cars is the marginalization of those who cannot afford one and cannot drive because of age or physical condition. As cars sweep other forms of travel aside, those without find it ever more difficult to gain access to jobs and services. Inequity of access is more troubling than aesthetic degradation. It undermines the very basis of a democratic society.

The solution to these various social concerns is not, however, to displace the auto. That would be a step backward. A more positive approach is to increase access for everyone, by improving public transit, by making cars that are easier to drive, and by broadening the use of telecommunications.[14]

More Than a Love Affair

In deciding what to do with the auto, it is critical to compare the overall costs and benefits. So far we have focused only on costs, which are substantial. But they are overwhelmed by the social benefits of driving. The fact is that people value cars. They buy cars as soon as they can afford them, and they prefer them to other means of transport. This attachment is not the "love affair" suggested by advertisements; it is not based in sensuality. The attachment derives from the unprecedented freedom, privacy, convenience, and security that cars provide.

Two reputable studies agree that the advantages of cars far exceed the costs, even when all unpaid social costs are included. Delucchi, in the Office of Technology Assessment study, estimated the benefits to be twice as large as the total cost to society. A leading environmental group, the Environmental Defense Fund, concurs: it estimated the benefits in southern California to be 60 percent greater than the total paid and unpaid costs.[15] Both studies measured benefits by examining the willingness of travelers to pay for auto travel.

Rising car sales provide hard evidence of the perceived benefits of motor vehicles. Even in the United States, with more than one vehicle per licensed driver, ownership is still expanding. Other economically advanced countries are not far behind; car markets are nowhere near saturation.

Increases are even more dramatic in less affluent countries, where the vast majority of the world's population lives.[16] From 1965 to 1987, auto ownership per capita increased twentyfold in China to 0.09 per hundred,

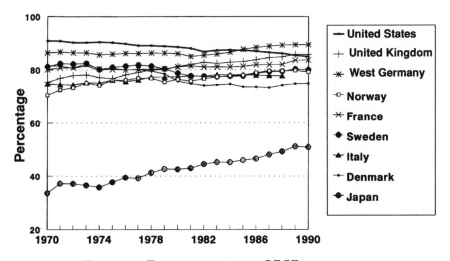

Figure 1-1 TRAVEL BY AUTO IN OECD COUNTRIES.
Source: Lee Schipper and Steve Myers, *Energy Efficiency and Human Activity* (Cambridge, England: Cambridge University Press, 1993).

and tenfold in the former Soviet Union, to 4.5 per hundred. In three European countries—Portugal, Greece, and the former Yugoslavia—car ownership increased almost tenfold to 13.1 per hundred. Based on evidence that oil used for transportation in developing countries increased 5.8 percent per year from 1970 to 1987, a period when vehicles were becoming more energy efficient, one concludes that vehicle ownership in those countries was increasing at much more than 6 percent a year.

In most affluent countries, automobiles already meet 75 to 85 percent of domestic travel requirements.[17] Among countries represented in Figure 1-1, only in Japan do cars account for less than two-thirds of travel. Even there, where distances are short, congestion pervasive, and rail transit superb, cars are used for almost 60 percent of travel, and that figure is rising rapidly.

Taming Drivers

One way to reduce the social costs of cars is to suppress their use—by inducing people to ride transit or bicycles, to telecommute, share rides, change workplaces or residences, and shorten or eliminate motorized trips. Strategems for encouraging such behavioral change fall into two categories: mandates and incentives. Neither has been very successful.

The legal premise for curbing solo car travel in the United States has been air pollution reduction. When metropolitan areas fail to meet air quality standards—as is the case almost everywhere—local governments are under a deadline to devise and implement plans to attain those standards. Driving is a central variable in such plans. To date, the so-called transportation control measures (TCMs) aimed at reining in car use have been remarkably unsuccesful. A 1993 U.S. General Accounting Office review concludes, "Virtually none of the literature we reviewed or the persons we interviewed stated that TCMs would significantly reduce emissions."[18]

It is not for lack of effort.[19] The largest and most aggressive such strategy attempted in the United States, known as Regulation XV, was adopted in December of 1987 in the Los Angeles area. Regulation XV is designed to reduce the number of vehicle miles traveled and trips taken between home and workplace. It requires employers of 100 or more individuals—more than 8,900 businesses—to prepare and implement plans for reaching this goal. The plans must include credible initiatives to encourage workers to share rides or use public transit. The overall target is 1.5 occupants per vehicle.

Based on a detailed study of 1,110 participating businesses, implementation of Regulation XV reduced work trips by about 3 percent during the first year (see Table 1-1). As a result, average vehicle occupancy went from about 1.22 persons per trip to 1.25.[20] A similarly small increase was measured in the second year. Interestingly, the rise came from increased use not of public transit but of carpools.

Because Regulation XV covers only work trips to large facilities, its impact was even smaller than these figures suggest. That is because facilities with 100 or more employees account for only 40 percent of all work trips, and work trips account for only 25 percent of total daily trips.[21] Assuming Regulation XV could reduce trips to these larger facilities by 5 to 10 percent, the net reduction in total daily trips is only about 0.5 to 1 percent.

These small gains came at large expense—about $3,000 per vehicle removed from peak hour traffic.[22] The gains come from those most responsive to incentives—commuters to large employment sites. The program may have been cost-effective in skimming these few motorists out of their cars, but further gains would require more incentives with a stiffer price tag.

Those driving to large facilities are the easiest to divert to alternatives—that is, carpooling or mass transit—because these facilities are the destination of many people. The meager gains of Regulation XV are therefore about as promising as can be expected from mandated reduction—especially considering how unpopular it is. An article in the San

TABLE 1-1
Impact of a Trip-Reduction Ordinance on Car Use (Regulation XV in the Los Angeles Area)

Mode of Transportation	Pre-Reg XV Mode Share	Post-Reg XV Mode Share	Change in Mode Share
Solo driving	0.757	0.709	-4.8
Carpool	0.138	0.184	+4.6
Vanpool	0.021	0.024	+0.3
Bus	0.032	0.032	0.0
Walk/bike	0.029	0.028	-0.1
Telecommuting	0.006	0.005	-0.1
Compressed work hours	0.016	0.019	+0.3

Source: G. Giuliano, K. Hwang, and M. Wachs, "Evaluation of a Mandatory Transportation Demand Management Program in Southern California," *Transportation Research* 27A(2): 125–38 (1993).
Note: Data in this table are for 1,110 workplaces.

Francisco *Chronicle* on November 22, 1992 described a similar proposal for the Bay Area as "a costly and controversial experiment in behavior modification, with 1.2 million commuters as the guinea pigs and the region's major employers picking up the tab."

The difficulty of discouraging people from driving solo is further supported by detailed studies of a larger range of incentive-based demand reduction strategies. A sophisticated analysis of the Los Angeles area found that a tax of $3 per day on parking would reduce travel by only 1.5 percent; a $110 annual vehicle registration tax, only 0.4 percent; and free transit for all families with incomes less than $25,000 per year, only 2 percent.[23] A large gasoline tax of $2 per gallon was estimated to have a larger effect, an 8.1 percent reduction in travel, but a tax of that size is politically implausible anytime soon. Moreover, as we will see, energy use and emissions can be reduced much more than 8 percent with any of a number of technological options.

The Shrinking Role of Mass Transit

The prospects for mass transit as a solution are even grimmer. Although public subsidies for transit have increased dramatically in the United States,[24] the number of workers commuting by public transit decreased from 8.4 percent in 1969 to only 5.5 percent in 1990, and those carpooling to work (defined as two or more occupants per car) decreased from 19.7

percent in 1980 to 13.4 percent in 1990.[25] Those driving alone increased from 64.4 to 73.2 percent between 1980 and 1990.

Mass transit is playing a shrinking role in virtually all economically advanced countries. In the United States, it accounts for only 4 percent of all passenger travel. In Western Europe, it has spiraled below 20 percent in most countries.[26] The gap between the United States and other affluent countries is continuing to shrink as motor vehicles become more common in those other countries.

Transit works best where there are large numbers of trips originating or terminating in a small, concentrated area, such as a downtown, a college campus, or a large industrial park. In dense cities such as New York and Tokyo, cars are costly and inconvenient, and transit is the preferred mode.[27] Theoretically, creating the same inconveniences and costs in more locales would induce more people to abandon their cars, but that seems neither desirable nor likely.

Even if mass transit use were expanded, thereby cutting down on the number of solo car trips, the energy savings and pollution reduction would be minimal. That is because transit use in the United States is so low to start with, and because transit consumes about the same energy per rider as cars.[28] Even a herculean doubling of transit ridership would lessen vehicle trips in the United States by only 4 percent, resulting in much less than a 4 percent reduction in emissions and energy use. A small reduction such as that could be gained much more easily and cheaply by tinkering with car technology.

It is almost impossible to imagine a scenario in which public transit as we know it in the United States would significantly expand its role in passenger travel. Conventional transit is ill suited to contemporary suburban land use patterns, and all trends—the information revolution, affluence, the wide availability of private vehicles, rising population, and the continuing decay of inner cities—suggest that suburbanization is unlikely to be reversed.

Toward a Technical Fix

If mandates, incentives, mass transit, and other strategies to reduce vehicle travel show little promise, then what can be done to reduce the large social costs of motor vehicles? The answer, the focus of this book, is founded on technical fixes. Technical fixes preserve the fundamental attractions of vehicle travel—mobility, convenience, and privacy—while requiring few behavioral changes. They support rather than subvert travelers' wishes and needs. Given the shortcomings of travel reduction strategies, and the huge promise of new technologies, the focus of any effort to create a more environmentally benign transportation system should be technical innovation.

The public strongly prefers this approach over restrictions on their behavior. In a 1991 survey conducted in the Los Angeles area, over half the respondents (57 percent) expressed willingness to purchase an alternative fuel vehicle as a response to air pollution problems, compared with only 17 percent who were willing to carpool, 16 percent who would use mass transit, and 6 percent who would walk or bicycle.[29] Although the survey covered only 230 individuals, the majority opinion was clear and forceful: travelers value their vehicles very highly. In the greater scheme of things, the behavioral changes to accommodate the inconveniences associated with electric and other alternative fuel vehicles are trivial compared to the changes associated with switching to mass transit and even carpools.

Some technical fixes have been highly effective in improving energy efficiency and limiting pollution. Increasingly sophisticated emission control technology, incorporating advanced electronics and combustion designs, have brought motor vehicle emissions down by as much as 90 percent for some pollutants. The cost has not been cheap, though—almost $800 per vehicle in 1990.[30]

Still greater emission reductions are possible and likely. In 1990, the California Air Resources Board adopted a new set of vehicle emission standards that are the most stringent in the world, requiring as much as 80 percent additional reduction in new-car emissions between 1992 and 2003. On February 1, 1994, the northeastern states as a group requested permission from the U.S. Environmental Protection Agency (EPA) to adopt the California standards for themselves. Approval is expected in late 1994.

Technical fixes have also substantially improved the energy efficiency of vehicles. In 1990, new automobiles in the United States used only about half as much energy as they did in the early 1970s. Most of the improvement came from more efficient engines, improved aerodynamic designs, lighter weight materials, and other relatively unobtrusive technical changes.[31]

These energy and emissions improvements, impressive as they are, don't go far enough. Despite emission mandates that reach into the next century, air will be unhealthy in many regions and still fall short of meeting ambient air quality standards; cars and trucks will consume more petroleum; and ever greater quantities of greenhouse gases will be released into the atmosphere.

What is the answer, if behavioral change strategies fail and incremental technical fixes are limited? Simple: variety and synergistic combinations. Combining behavioral strategies with technology strategies enables the pursuit of much more ambitious technologies (and attainment of greater behavioral changes); together they would provide benefits that dwarf what is possible with any single plan or mandate. The challenge is to identify superior technologies that yield long-term, sustainable advantages, and to design pricing and regulatory strategies that support change. The following presents a chronology of what is possible—of what actions and events might lead to the creation of an environmentally benign transportation future.

Imagining the Future: An Optimistic Chronology of Events from 1995 to 2020

The following chronology is a vision of possibility—the creation of an environmentally benign transportation system in the not so distant future. Underlying this optimistic vision is the belief that sooner or later some form of electric driveline will begin to replace internal combustion engines, that smaller vehicles will eventually replace larger vehicles in certain applications, and that today's fragmented regulatory approach will become more flexible, coherent, and incentive based.

1995

President Clinton renews his 1993 pledge to build superclean and efficient cars. The pledge is backed by increased funding for innovative technology companies. Large automakers commit to specific production and performance targets for improved fuel economy and emissions in return for assistance with research and development.

1996

Revenue-neutral fee-bate programs to tax polluting vehicles and to provide rebates to clean vehicles are adopted in California and several other states, reducing the cost of electric vehicles to consumers by an average of $3,000 per car.

1997

Tradeable greenhouse-gas-emission standards are adopted by the federal government for all cars and light trucks, followed in California by more sophisticated air pollutant and greenhouse-gas-emission trading programs.

This further reduces the cost of new electric vehicles, by up to $2,000 per car. Automakers shift their $5 billion plus research and development budget for controlling gasoline-car emissions and efficiency to electric vehicles, mostly zero-emitting fuel cell vehicles.

1998

A small number of hybrid vehicles and buses retrofitted with fuel cells are introduced. The fuel cells run on hydrogen reformed from natural gas.

1999

Two new towns being built in California and Florida make a combined purchase of 20,000 very small neighborhood electric cars, to be given away free with the purchase of a home. Streets are specifically designed for neighborhood cars. House sales exceed expectations; people are willing to pay premiums of up to $40,000 per house.

2000

Ford offers eight free days of car rental each year for four years with every new purchase of an electric vehicle. General Motors, Toyota, Nissan, Honda, and Chrysler match the offer. The cost to the manufacturer is small, equivalent to a $600 rebate.

Southern California Edison and Boston Edison create subsidiaries to market electric vehicles. Hertz soon follows suit. These "mobility providers" rent cars to consumers for a monthly fee. They assume all maintenance, registration, and insurance responsibilities, and will upgrade or exchange vehicles on two weeks notice.

2001

Santa Monica, Berkeley, Davis, and Palm Desert, California, ban all full-size vehicles from 9 A.M. to 4 P.M. in selected neighborhoods. Many communities follow suit. Speed limits of 20 mph are set on many streets.

2002

U.S. Electricar and a major manufacturer of composite materials join forces to manufacture lightweight electric vehicles. An entirely new manufacturing process is created, one suited to customized vehicles and not dependent on large economies of scale.

2003

The first true fuel-cell bus enters commercial production. Transit operators are enthusiastic about its long range and zero emissions.

2005

Semiautomated controls are made available as an option on neighborhood cars. These allow elderly persons and others with minor physical disabilities to operate cars.

2008

The first mass-produced fuel-cell cars are sold. Battery-powered vehicles account for 25 percent of new cars and trucks sold in California; one-third of these are neighborhood electric vehicles.

2111

Fuel-cell vehicles account for 10 percent of vehicle sales in California, and 5 percent elsewhere in the United States.

2013

Construction of a solar-hydrogen energy farm begins in the southern California desert. The project is a joint venture of Southern California Edison, Southern California Gas, and Arco. Other oil, natural gas, and electricity companies soon invest in solar-hydrogen farms to capture the growing fuel-cell-vehicle market.

2020

Most new vehicles in California are now zero emission. Fuel cells by themselves account for half of all vehicle sales. The use of fossil fuel for transportation is steadily declining. Only 15 percent of new light-duty vehicles in California, and 30 percent elsewhere in the country, are fueled directly by natural gas, methanol, or gasoline. Greenhouse gas emissions associated with new vehicles are one-third what they were thirty years earlier.

CHAPTER 2

Sifting the Wheat from the Chaff

Government policy toward motor vehicles is fragmented and increasingly misguided, resulting in small environmental benefits being gained at exorbitant cost. In response to government requirements, U.S. oil companies are lavishing billions on reformulating gasoline, and automakers are spending a similar amount to redesign gasoline engines for lower emissions. One large grain company is receiving nearly a half billion dollars annually in subsidies to make ethanol fuel from corn, and U.S. consumers are spending at least that amount to test the emissions from their cars.

Meanwhile, legislators and bureaucrats have created a complex tangle of rules tying transportation investments to air quality. The air quality tail is wagging the transportation dog. That's not right. It ignores the other costs of transportation—including energy and greenhouse effects—as well as transportation's benefits. Even more troubling is that many of the rules are ineffectual. This is not to say that all government efforts to create a more environmentally benign transportation system have failed—many have been stunning successes. But what worked in the past will not necessarily work in the future.

We are at a crossroads in the transportation sector. The recent outpouring of technological initiatives and stepped-up experimentation with new regulatory approaches and forms of governance are evidence of changing circumstances. Some of those initiatives make sense; others don't. Some would make more sense if redirected. The net effect is disappointing. The current incrementalist path shows little promise of moving the transportation system to an environmentally sustainable future.

How do the lessons of the past inform the initiatives of the present?

Sensible but Limited Options

Many current initiatives are reasonable but limited. In this category are those that made more sense in the past than they will in the future, and those that were ineffective or ignored in the past and show some promise for the future. But of those addressed below, none hold much long-term potential for creating a more benign transportation system.

Tracking Down Super-emitters

One increasingly ineffective and costly strategy is mandatory testing of vehicle emissions.[1] The goal of mandatory testing, required at least biennially in all regions violating air quality standards, is to identify "super-emitters"—those cars that emit much more now than when they were new. Emissions from internal combustion engines can deteriorate precipitously, sometimes increasing over a hundred-fold after only a few years. Super-emitters are those 10 percent or so of cars and light trucks that account for half of the emissions.[2] (The cleanest 50 percent produce less than 1 percent of emissions.) Super-emitters tend to be older vehicles, but some are relatively new vehicles with poorly maintained engines, defective emission-control equipment, or emission controls that have been tampered with.

In the 1980s, inspection and maintenance programs brought emissions down by an estimated 12 percent. That trend has ended. The problem in most states is that inspections are carried out at small neighborhood service stations, an expensive and increasingly ineffective system. Small stations have little incentive to be diligent in conducting the tests—many are not willing to risk offending customers and losing future business—and they cannot afford to buy the sophisticated, accurate testing machines that cost hundreds of thousands of dollars. As a result, regulators have been forced to accept the use of inferior testing equipment. The net result is a failure to eliminate emissions from the 10 percent who are super-emitting.

The best solution appears to be greater use of unobtrusive, remote-sensing devices situated next to roadways. These would detect passing super-emitters and automatically photograph license plates. A notice would be sent requiring the owner to bring the vehicle to a large testing station with accurate (and much more expensive) state-of-the-art testing devices. In principle, only those vehicles likely to be in violation would be forced to undergo the expense and inconvenience of an inspection.[3]

Remote sensing is not a panacea. The devices are not always accurate. They have difficulty measuring emissions from vehicles in interior lanes, as well as distinguishing between genuine super-emitters and vehicles that are merely cold or accelerating. (Emissions are many times higher when the engine is cold and increase manyfold during acceleration.) As a result of these shortcomings, some car owners would be subjected to the inspection process without good cause. The cost and inconvenience to individuals would be much greater than now, because the tests would be more expensive and time-consuming and available at fewer and therefore more remote stations. A partial but expensive solution to this false-detection problem would be to place police officers downstream of the remote-sensing devices. They would flag down suspected super-emitters to confirm through visual inspection whether the vehicle had high emissions or had been tampered with.

The cost of creating a network of remote-sensing devices would not be trivial. To reduce the possibility of apprehending nonsuper-emitters, and to assure good coverage of a region, many devices would need to be erected, and they would have to be operated for extended periods.

Considerable political effort and expense have gone into creating and updating the existing vehicle inspection and maintenance program. The EPA has tried to require centralized testing for quality control, but it has resisted efforts to use remote-sensing devices. Meanwhile several states, led by California, have insisted on maintaining a decentralized inspection program. Without a sharp break from current practice, it is probably wishful thinking to expect more than modest improvements in emission reduction through testing.

Cash for Clunkers

Cash for clunkers is a program endorsed in the 1990 Clean Air Act for getting super-emitters off the roads altogether. Typically, it would function by offering a cash payment to anyone junking a vehicle older than a specified model year.[4] The cash would usually come from a company looking for pollution credits. The company buys credits in lieu of reducing pollution from its factory, refinery, or bakery, or to enable it to expand its pollution-emitting production facilities. This limited form of credit trading—between scrapped cars and stationary facilities—has the economic benefit of reducing pollution at less cost than otherwise, and the political benefit of making it easier to enact more stringent pollution rules. The program has been strongly supported by automakers as a way to hasten vehicle turnover and therefore sales of new cars, by companies seeking relief from inflexible air quality rules, and by free-market ideologues attracted to market-based rules and incentives.

Again, reality falls short of promise. How does one screen out the many old vehicles that are not super-emiters, and those that were going to be junked soon anyway? Cash for clunkers could prove to be a costly method for reducing pollution if appropriate procedures and tests are not enacted. At present, all ideas for implementing such procedures and tests appear expensive. Pilot projects have been conducted in Delaware, Illinois, and Los Angeles,[5] but because of unresolved uncertainties—how much the vehicles were being driven, their actual emissions, how much they would have been driven if not scrapped, and with what they were replaced—a definitive evaluation of their effectiveness and cost cannot be made.[6]

Even assuming the creation of cost-effective screening, total emission reductions would be modest because very old cars are generally not driven much anyway, many super-emitters are not old enough to be captured by these scrapping programs, and such programs tend to have a one-time appeal. Their luster will dim after the worst offenders are junked.

Uniform Emission Standards

As indicated earlier, uniform new-car emission standards have been highly effective. By requiring every vehicle to meet the same performance standards, regulators succeeded in gaining huge reductions in emissions. New cars in the early 1990s emitted only about one-fourth as much pollution as uncontrolled vehicles of the 1960s.[7]

Their appropriateness for the future, however, is less certain. The cost in the early 1990s for complying with uniform emission standards was about $700 to $1,000 per vehicle.[8] Further emission reductions will require more complex and expensive control technology. A more fundamental concern is that continued use of uniform standards robs automotive engineers and marketers of flexibility. It forces them to treat every vehicle and sales region equally. It forces them to ignore the fact that emission control costs are lower for some types of vehicles, engines, and fuels than others, that certain pollutants are of greater concern in some regions than in others, and that uniform emission standards may hinder the attainment of other social goals, such as energy and greenhouse gas reductions. The net effect of these flaws is a terribly inefficient system. The current approach of uniform standards for all vehicles blocks innovations such as lean-burn combustion designs and two-stroke engines (virtually all of today's car engines are four stroke). These innovations, which have the potential to bring substantial savings in energy and pollution, are virtually precluded from the market because they cannot easily meet the uniform emission standard for one category of pollutants—nitrogen oxides. Other innovations, such as electric propulsion, would yield even greater benefits than lean-burn and two-stroke engines, but uniform standards by their nature eliminate industry's incentive to better the standards, and for consumers to buy cleaner-burning vehicles. With uniform standards, consumers get no financial reward for buying cleaner cars and industry gets no reward for making them. The current system does not encourage, and in some cases discourages, innovation and flexibility.

CAFE

Corporate average fuel economy (CAFE) is another sensible and effective regulatory initiative that has outlived its usefulness. CAFE standards were established by Congress in 1975 and took effect in 1978. They require that the average fuel consumption of all cars and light trucks sold by a manufacturer in a given year not exceed a specified level. The standard for cars was 18 mpg in 1978 and gradually tightened to 27.5 mpg in 1985, where it still stands. CAFE standards were effective in lowering the fuel consumption of new American cars during those years. David Greene, a respected analyst from Oak Ridge National Laboratory, has shown that they were twice as effective as high gasoline prices in improving fuel economy.[9] The automotive industry, stridently opposed to CAFE standards, disagrees with the findings, giving high gasoline prices the credit.[10]

In any case, since the mid-1980s, the fuel economy of vehicles—fuel consumed per mile of vehicle travel—has stagnated and even worsened (see Figure 2-1). What has improved is fuel *efficiency*—the measure of how much initial fuel is converted into usable energy. But greater efficiency has been offset by the use of larger, more powerful engines and by such energy-consuming features as all-wheel drive and air conditioners. Worse, the light truck, which has much poorer fuel economy than the car, has gained an increasing share of the vehicle market. Separate fuel-economy standards exist for the light truck, but they edged up by only 0.1 miles per gallon per year through the early 1990s, not even reaching 21 mpg for model year 1996. The stagnant CAFE standards for cars and virtually stagnant standards for light trucks are the direct result of the automotive industry's virulent opposition to any increases.

U.S. cars now have nearly the same fuel economy as European cars, even though fuel prices are several times higher in Europe and compact European cities are inhospitable to big cars (see Figure 2-1). If Europe's tripled fuel prices fail to induce much greater fuel economy, what will? Given the small likelihood of major fuel price increases in the United States and automotive industry hostility to increased CAFE standards, only small improvements in fuel economy can be expected—probably no more than 1 to 2 percent a year for the next two decades[11]—unless entirely new and inherently more energy efficient vehicle technologies are developed and introduced.

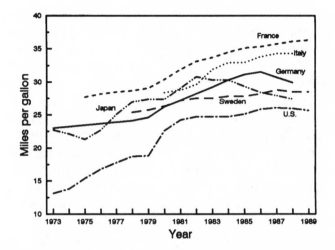

Figure 2-1 Fuel economy of new cars in OECD countries.

Source: Oak Ridge National Laboratory, *Transportation Energy Data Book,* 13th ed. (Springfield, Virginia: NTIS, 1993). Based on analysis by Lee Schipper.

Note: Diesel cars are excluded in West Germany, where they comprise a large portion of total light-duty vehicle sales. Light trucks, which make up a large portion of light-duty vehicle sales only in the United States, are included. These fuel economy ratings are test values; actual on-road fuel consumption is higher.

Raising Prices to Restrict Auto Use

In the previous chapter, it was shown that major reductions in vehicle use are not likely nor desirable. People want more mobility and accessibility, not less, and are willing to pay for it. But not all trips are indispensable and highly valued. A late night trip to purchase milk or a six-pack could be eliminated—and would be—if the cost of that trip were high enough. Likewise, some people would switch from solo car travel to mass transit and carpools if the cost of driving or parking were raised. Some travel reduction is possible and desirable.

One approach gaining favor among analysts and decision makers is, as suggested above, to raise the price of driving to reflect the true costs of driving. Pricing mechanisms are considered more acceptable than outright restrictions on behavior because they offer the consumer a choice. Better pricing is a sensible option that has been mostly ignored, but the potential effects are limited

In the economist's ideal world, correct price signals would be the sole requirement for reducing the use of polluting and fuel-guzzling vehicles. In the real world, however, raising prices is politically anathema. Voters and companies would fight government efforts to impose higher road

taxes if they perceived no alternative to driving, and they would fight taxes on polluting cars if low-polluting cars were unavailable. Opposition would mount in proportion to the size of the tax. By themselves, taxes of the scale needed to change behavior will face almost certain political death. Pricing cannot be relied on as the principal means of changing driving habits.

The overriding lesson, as suggested earlier, is that pricing and other regulatory strategies need to be employed together with technology and infrastructure options that expand choice. A combined strategy will be far more effective than any single strategy. For example, a tax on gas guzzlers could be paired with subsidies for the development of new vehicle and energy technologies. The tax would influence people and companies to select the more socially beneficial alternatives made possible by technology development subsidies.

Smart (and Quasi-Smart) Transportation

Traffic congestion is getting worse every year, especially in suburban areas, and yet few new highways are likely to be built in the United States. One solution, as suggested above, is to attach a price tag to roadway and vehicle use that fully reflects its cost to society; that would certainly result in less travel and less congestion. Another possible solution far more attractive to transportation officials and politicians is the development and deployment of what has become known as intelligent vehicle and highway system (IVHS) or intelligent transportation system technologies.

With the interstate highway construction program begun in 1956 nearing completion, the highway engineering community is avidly seeking a new vision and mission. IVHS is what it has settled on. Geared primarily to getting more use out of existing roadways, IVHS is now being tried in a variety of publicly funded demonstration projects around the United States (and in Europe and Japan). A recent U.S. Department of Transportation plan guiding IVHS research suggests that "over the next 20 years, a national IVHS program could have a greater societal impact than even the Interstate Highway System."[12] What will the impact be?

IVHS technologies attracting the most attention and resources are those that deliver information to drivers and metropolitan traffic managers. The objective is to deliver accurate up-to-the-minute traffic information so drivers can select the fastest or easiest route, and find parking, restaurants, and other destinations more easily. This information is expected to be most valuable in helping drivers avoid traffic jams, thereby reducing congestion. Analytical studies, however, find that traffic and route information rarely saves a driver more than a few minutes a day, because there are usually no acceptable alternative routes.[13] (Los Angeles, with its large network of parallel freeways and arterials, may be an exception.)

Traffic information, by reducing congested stop-and-go traffic, could also reduce pollution. That is because a large proportion of pollution is emitted during the hard accelerations common in stop-and-go traffic. One pedal-to-the-floor acceleration of a few seconds can generate as much exhaust as the entire remainder of a trip.[14,15] The energy savings from smoother traffic flow are likely to be less pronounced, as hard acceleration has a much smaller effect on energy use than on emissions.

The second major set of IVHS technologies is more visionary: automation of vehicle controls, that is, computerizing controls so that vehicles react more quickly and accurately than humans. This would not only make driving safer, it would also allow a narrowing of lanes and less following distance between cars, thereby expanding the capacity of existing roads. By some estimates, automated controls would more than double the capacity of roads and sharply reduce congestion.

That is the theory of automated highways. But vehicle automation faces a host of obstacles, including cost, liability, consumer resistance, and questions of equity.[16] One major challenge is how to maintain funding momentum, political support, and industry interest over the decades needed to reach full automation. During this extended period the benefits would be small.[17] Why? First, any vehicle not automated would not be closely following the vehicle in front of it and would therefore slow all traffic behind it. Second, until all vehicles were automated, lanes could not be narrowed. Third, until all lanes are automated, negotiating between the fast and slow lanes will be difficult and dangerous. This lane-switching problem, along with the fact that most trips on urban freeways—the type of roadway easiest to automate—are only a few miles long, makes it unlikely that automated lanes would be heavily used during the transition period. Fourth, the cost of building separate access and exit ramps to segregated automated lanes to eliminate the lane-switching problem would be extraordinarily expensive—hundreds of millions of dollars for each set of ramps. Would it be politically possible to invest hundreds of billions of dollars over decades before much benefit was seen?

Another concern is environmental.[18] On the one hand, stop-and-go movement and hard accelerations would likely diminish on automated highways, a substantial environmental benefit. On the other hand, the increased capacity and lesser congestion of automated highways would encourage more travel.[19] People would move to the fringes of metropolitan areas and drive longer distances, their cars using more energy and emitting more pollutants—factors that have led environmental activists to oppose automated control technologies.[20] One partial solution to these concerns would be to convert vehicles to electric propulsion.

Financial equity is another troubling concern. Who pays for the large infrastructure? Should all drivers pay for automated highways to support those fortunate enough to afford the expensive information and automated controls? In a greater sense, should government be encouraging the creation of a transportation system that marginalizes those less fortunate, especially at a time when economic forces are wiping out middle-class factory jobs and expanding the gap between the rich and poor?

Another obstacle concerns liability. Who would be responsible for an automated car that caused a chain reaction of collisions? If courts decide that manufacturers are liable, then companies would be reluctant to supply the technologies.

There is also doubt as to whether drivers would willingly forego control of their vehicles.

In summary, it is likely that information-based IVHS technologies will be attractive to motorists, as would be limited versions of automated technology, such as those that warn of collisions and assist partially disabled people. Inevitably, then, a profitable IVHS industry will take root with or without government subsidies. But the net effect will be relatively limited social benefit and little or no progress toward the creation of an environmentally benign transportation system.

Beyond Smart Cars

We can do better. An IVHS focused on the real goal of transportation—access to goods and services—rather than just more mobility, and shaped to be more responsive to goals of equity and environmental quality, could become a powerful tool for improving the twenty-first century urban and social landscape.[21]

Government would need to be more active in supporting products and activities that benefit lower-income classes and the environment. Technologies such as "smart" teleshopping, neighborhood cars, and electronic speed controls, largely ignored by IVHS proponents, would have to be embraced, and others such as smart paratransit, now under the IVHS umbrella, would have to receive more attention.

Shopping through interactive television and the creation of other smart information systems—for example, creating more and better information for the movement of goods and for smaller business inventory management—might halt the trend toward long shopping trips to regional warehouse stores.

Electronic speed controls could be incorporated on a variety of roads: on residential and low-volume roads to make them safer and to encourage travel by nonmotorized and small neighborhood vehicles; on arterials to smooth traffic flow and thereby reduce emissions from gasoline-powered vehicles (and enhance neighborhood car safety); and on freeways to make travel speeds more uniform, thereby improving safety and reducing emissions. Speed controls could be made more acceptable by providing for manual overrides of the controls in emergencies and exempting emergency vehicles such as ambulances, fire engines, and police cars.

Smart paratransit—ridesharing in small transit vehicles using information technology—may prove to be the most important IVHS initiative of all.[22] Smart paratransit is a modern update and major improvement over the old dial-a-ride concept and the burgeoning airport shuttle van services. Up-to-the-minute service and traffic information would eliminate the need for reservations. Travelers could request rides by telephone, cellular phone, interactive television, modem-equipped computer, or public computer terminal. The call would be routed to small transit vehicles passing nearby or even to individuals willing to accept passengers.

Smart paratransit would dramatically improve access for people without cars, as well as solve the dilemma of suburbs being too dense for cars and not dense enough for buses and rail transit. By filling the gap between large transit vehicles and cars, smart paratransit would attack metropolitan congestion head on.

There are some kinks to work out, among them, structures for the sale of information to ride providers, how to allocate transit subsidies, how to license providers and set their fees, and how to ensure the safety and security of riders. Researchers are just beginning to grapple with these questions.[22] The answers are still unclear. What is clear is that smart paratransit has the potential to curb driving and lower overall transportation costs. This would lessen pollution, save energy, and relax the need for new highway infrastructure, including parking and other ancillary facilities. It would also widen options for those who couldn't afford a car and those physically unable to drive.

Redirected IVHS technologies could serve as catalysts for still more far-reaching and positive change. Low-speed neighborhood electric vehicles, for example, might improve access for older and less physically capable people by making it easier for them to drive. (It is not difficult to incorporate semi-automated driver assists such as collision avoidance, smart cruise control, and assisted steering into low-speed vehicles.) As a viable alternative to full-size cars, small, low-speed vehicles could also strengthen emphasis on neighborhood centers and nonmotorized travel. As we shall see in Chapter 4, the deployment of electric vehicles might set in motion a series of events that eventually transforms communities and road infrastructure.

IVHS technologies can also enable the adoption of creative and clever pricing strategies to target those trips and vehicles that are most costly— clever in the sense of being politically acceptable and not overly compromising of equity. IVHS technologies are already being tested for this purpose. One example is the use of inexpensive transponders attached to license plates or elsewhere to receive signals from small transmitters placed along roadways. If the idea takes root, these signals would indicate the cost of traveling at a given time and place; a transportation utility would then send a road-use bill once a month to the car owner. Variations of this concept will have to be tested to find the most politically and economically viable format. For instance, charges might be revenue neutral, so that those traveling in carpools, buses, or smart paratransit, or not traveling at all, would receive a rebate each month. Or a debit card could be used, much like an ATM bankcard, making charges against a checking account and thereby eliminating any threat to the privacy of the motorist (neither government nor business would have any record of the travel).

Additional examples of IVHS–enabling pricing strategies include fees on polluting cars, with rebates for zero-emission and neighborhood electric vehicles; the application of revenues to cross-subsidize various smart paratransit operations; and payoffs to local residents who allow tolls to be charged on their streets.

Would pricing of this nature to accomplish transportation efficiency place a larger burden on the poor? Not necessarily.[23] Explicit provisions could be made to accommodate those who are "priced off" the roads and to provide better access to those who are financially or physically disadvantaged.

In summary, IVHS could play an important role in the transition to a more environmentally benign transportation system, not least by facilitating the adoption of equitable and efficient pricing tools. But this more positive role requires redirection of IVHS initiatives.

Alternative Fuels: The Good, the Bad, and the Indifferent

The most attractive alternative fuels are natural gas and two alcohols, methanol and ethanol. Reformulated gasoline is sometimes included in this list, but strictly speaking it is an improvement over conventional gasoline, not an alternative. While none of these alternatives merit priority consideration in a long-range sustainable transportation plan, alternative fuels for internal combustion engines can yield substantial reductions in urban air pollution, and in some cases lessen greenhouse gas emissions, thereby acting as a bridge to hydrogen and fuel cells. The most powerful argument against these fuels is that because society can be expected to embrace a transition to a new fuel only once in a great while, pursuing alternative fuels may well divert attention from electric propulsion, which promises greater and more lasting benefit.

Mistakes have been made and continue to be made in the pursuit of alternative fuels; they have a history of creating unfulfilled expectations. Options have emerged and then faded for reasons of cost or, more recently, because of insufficient environmental benefit. The challenge is to settle on those fuels and investments that make the most sense and are most likely to succeed.

The list of sensible fuels does not include those made from coal and oil shale. As a direct result of the 1973 Arab oil embargo and subsequent price rises in the late 1970s and early 1980s, the United States and its oil industry attempted to create energy independence by developing domestic resources. Tens of billions of dollars were spent—a few billion by the government and much more by large corporations—to convert oil shale and coal, as well as less plentiful materials such as tar sands and peat, into liquid and gaseous fuels.[24] The money was wasted. Sprouting investments quickly died when feasibility studies, funded by the federal government's Synthetic Fuel Corporation, showed the cost of producing fuels from coal and oil shale in first-generation plants to be around $100 per oil-equivalent barrel.[25] These synthetic fuels are unlikely ever to be revived, not only because they are expensive, but also because of the substantial environmental damage sustained in extracting them and the large amount of greenhouse gases emitted during production and combustion.

Reformulated Gasoline

Gasoline that has been modified to emit fewer hydrocarbons and less benzene and other pollutants is known as reformulated gasoline. It provides a modest improvement in pollution at a small cost. By the mid- to late 1990s, reformulated gasoline will replace conventional gasoline in much if not most of the United States, becoming the new "conventional" gasoline.

The success of reformulated gasoline comes mainly from strong government support. Shortly after President Bush proposed the sale of alternative fuel in the nine most polluted cities of the country in July of 1989, Atlantic Richfield (Arco), a medium-sized oil company whose prime market is in southern California, introduced EC-1, a gasoline reformulated to reduce hydrocarbon emissions.

Reformulated gasoline became a requirement in the fall of 1990 when President Bush signed a bill that amended the Clean Air Act and required reformulated gasoline in the more polluted cities of the country. Soon afterward, the California Air Resources Board went a step further and required that gasoline sold in the state be even more drastically reformulated beginning in 1996. The California and federal rules are expected to result in reductions in ozone pollution of up to 30 percent per vehicle per mile over conventional gasoline, while altering—not necessarily lowering—the levels of other pollutants. This new gasoline will likely add up to 15 cents to the cost of a gallon of gasoline.[26]

The oil industry has apprehensively embraced reformulated gasoline, intent on undermining government mandates for what to them is a greater evil: alternative fuels.

Methanol from Natural Gas

Methanol, a liquid, is usually made from natural gas in large petrochemical refineries. It can be made at greater cost from coal and biomass (plant material). Methanol, though enthusiastically promoted by segments of some California and federal agencies, has now been shown to be of little advantage.

The most recent evidence suggests that methanol's use in internal combustion engines could have a small positive effect on urban ozone pollution.[27] This wouldn't be the case if methanol were burned at 100 percent strength—then the gains would prove substantial—but safety considerations and cold-start problems probably rule out that possibility. A combination of 85 percent methanol and 15 percent gasoline, known as M85, would benefit air quality by much less than M100. The oil industry and others have claimed that vehicles running on M85 would not be any cleaner than those run on reformulated gasoline. That may be overstating the case, but it's not far off. Flexible-fueled vehicles that allow any mix of methanol and gasoline—the only version of methanol cars contemplated at present—shrink the benefits even further. Indeed, according to a prestigious National Academy of Sciences study, "If most vehicles were running on a more dilute blend (say M50 or M25), increased evaporative and other organic emissions could lead to increases in ozone."[28]

Methanol (made from natural gas) is even less attractive as a greenhouse gas reduction strategy. The latest and most detailed studies suggest that methanol-powered light-duty vehicles will result in about the same amount of greenhouse gas being emitted as gasoline-powered vehicles—taking into consideration all greenhouse gases and all upstream sources of the gases (see Table 2-1).

While the environmental benefits of methanol are small, so are the additional costs. Estimates are that methanol made from imported natural gas would cost about 20 cents more per gallon than regular unleaded gasoline.[29] This is less than 3 percent of the cost of owning and operating a vehicle. The cost would be a little higher if the methanol were produced from coal, biomass, or U.S. domestic natural gas.

The auto industry has been relatively accepting of methanol because it creates few risks or additional costs for them. Fuel-flexible vehicles can burn any combination of methanol and gasoline, and they cost about the same as conventional gasoline vehicles when mass-produced—and only a few hundred dollars extra with limited production runs. In a worst-case scenario, automakers could simply sell fuel-flexible methanol vehicles as gasoline vehicles; the driver need never know the vehicle could also operate on methanol, thereby bypassing any consumer resistance.

Even with this minimal risk auto industry investments in methanol continue to lag, presumably because automakers view methanol as not worth the trouble. They are still selling a small number of fuel-flexible methanol vehicles, while reportedly investing virtually no resources in the design and construction of very-low-emitting methanol vehicles. As already noted, the industry seems to prefer meeting tougher emissions standards in California and elsewhere by improving gasoline combustion and emission controls.

The oil industry's investment in methanol would be larger and riskier than the auto industry's. To compete in a methanol market, the fuel industry would have to invest a billion or more dollars per methanol plant as well as a more modest sum to deliver and sell fuel. Like the oil industry, the petrochemical industry is skeptical about this fuel. This skepticism became apparent at the industry's 1987 World Methanol Congress held in San Francisco. At that meeting, California politicians and administrators made a strong pitch for using methanol as a transport fuel. But in so

TABLE 2-1

Change in Greenhouse-Gas Emissions from Gasoline-Powered to Alternative-Fuel Vehicles

Fuel/Feedstock	Change (percent)
Fuel cells, solar-powered hydrogen	−90 to −85
Ethanol from wood	−75 to −40
Electric vehicles, natural-gas plants	−50 to −25
CNG from natural gas	−20 to 0
Methanol from natural gas	−10 to +8
Electric vehicles, current U.S. power mix	−20 to 0
Gasoline	—
Electric vehicles, new coal plant	0 to +10
Ethanol from corn	−10 to +35
Methanol from coal	+30 to +70

Source: Mark A. DeLuchi, "Emissions of Greenhouse Gases from the Use of Transportation Fuels and Electricity" (Argonne National Laboratory, Center for Transportation Research, Argonne, Illinois, 1991, ANL/ESD/TM-22).

Notes: Methane, nitrous oxygen, and hydrocarbon emissions from vehicles are converted into the mass amount of carbon-dioxide emissions with the same temperature effect, defined as the same number of degree-years over a given time. This analysis considers emissions of methane, nitrous oxides, and carbon dioxide from the production and transportation of the primary resource (coal, natural gas, or crude oil), conversion of the primary resource to transportation energy (for example, natural gas to methanol), distribution of the fuel to retail outlets, combustion of the fuel in engines, and the manufacture of vehicles.

doing, they made it clear to industry executives that methanol would need to be price-competitive with gasoline—thereby sharply curtailing profit opportunities.

Methanol's ultimate success hinges on the development of more environmentally friendly production, perhaps using biomass or more efficient and benign propulsion technologies, such as fuel cells. The most outspoken advocates have been several government leaders: Charles Imbrecht, the appointed chair of the California Energy Commission; Charles Gray, a respected technical manager of research at the U.S. Environmental Protection Agency; and during the Reagan and Bush administrations, Boyden Gray (no relation to Charles Gray), a key White House advisor to George Bush. Remarkably, these few individuals were able to generate strong sustained government support for methanol from the mid-1980s through the early 1990s. Outside this circle, however, methanol has few enthusiastic proponents, and its future remains in doubt.

Alcohol Fuels from Biomass

Alcohol fuels derived from biomass crops such as trees represent the only nonelectric transportation option that could dramatically reduce greenhouse gas emissions. But its other environmental attributes are not so positive; especially troublesome are adverse impacts on biodiversity caused by growing and processing so much biomass—not to mention the vast tracts of land required to grow biomass.

There are essentially two biomass fuel options: fermenting starch and sugar crops into ethanol, or converting the cellulose of biomass, such as wood pulp, into ethanol, methanol, or hydrogen.[30] The second option appears less expensive and more environmentally attractive than the first.

Fermentation of farm crops to produce ethanol is already common in some parts of the world. In Brazil, the fuel is fermented from sugarcane and used in vehicles designed to run only on ethanol. Through the mid- and late 1980s, ethanol cars accounted for over 90 percent of all cars sold in the country; even in the 1990s, with low petroleum prices, ethanol cars still represented over half of all sales. The conditions for fermenting and distilling farm crops into ethanol are more favorable there than anywhere else in the world, mostly owing to the country's efficient sugarcane industry.[31]

In the United States, 5 to 8 percent of gasoline sold since 1980 has been a blend containing 10 percent ethanol made from corn. Sold mainly in midwestern farm states (but elsewhere as well), ethanol fuel costs more than twice as much to produce as gasoline. The industry exists only because of extravagant federal and state subsidies. In 1990, combined federal and state subsidies for ethanol amounted to $467 million.[32] The per-gallon subsidies more than equal the wholesale cost of a gallon of gasoline. The subsidies exist principally because of the influence of Archer Daniels Midland, a large food-processing and grain-trading company that has maintained close ties and provided substantial campaign contributions to many key congressional leaders and several recent presidents. ADM produces approximately two-thirds of all the fuel ethanol in the country.

The market for ethanol fuel is likely to expand in the mid-1990s as a result of intense lobbying to mandate that ethanol be an additive in reformulated gasoline. However, corn ethanol is undesirable both on economic and environmental grounds. When blended with gasoline and combusted in motor vehicles, corn ethanol reduces carbon monoxide output but increases other smog-causing emissions.[33] More important, growing corn for ethanol consumes a large quantity of fossil fuel—as an input to fertilizer and to power farm equipment and distillation boilers. Considering the full energy cycle, ethanol fuel made from corn currently generates about the same amount of greenhouse gas emissions as gasoline.[34]

The second biomass option—converting cellulose into ethanol or methanol—is more attractive. The prime biomass candidate is trees, because they are plentiful and relatively inexpensive. Other attractive biomass candidates include grass and residues such as paper and sawmill waste.

Because they can be grown on less fertile soil, have a high yield per acre, and require relatively little energy input, trees are preferable to corn, sugarcane, and other starch and sugar crops. The U.S. Department of Energy has set a cost goal of producing wood alcohol for less than one dollar per gasoline-equivalent gallon, but this ambitious goal has not yet been realized.[35]

The prinicipal drawback of wood-alcohol fuel is environmental damage caused by plantation-style farming. On most current tree plantations, single species are planted over vast areas and nurtured with irrigation, pesticide, and fertilizer. This monoculture style of farming has the advantage of reducing direct costs but the disadvantage of diminishing biodiversity.[36] Moreover, frequent harvesting, fertilizing, and spraying exacerbates erosion and runoff of chemicals and other pollutants into water supplies. These environmental problems are not inherent to intensive forest and farm management. More sustainable techniques such as planting of mixed species, reducing energy input, and harvesting less intrusively are possible. However, unless adverse environmental effects are figured into the cost of doing business, the adoption of more sustainable practices is likely to be slow at best. The poor record of land-use regulation and enforcement in the past does not inspire confidence for the future. If biomass is to become an important energy source, major changes in production will be needed to keep environmental damage to a minimum.[37]

Cellulosic biomass options are most attractive in regions where wood and grass are plentiful and can be grown and harvested inexpensively. Care must be taken, however, to assure that regulatory and policy frameworks nurture positive biomass options and squelch negative ones. The corn ethanol experience illustrates how good ideas can be distorted by vested economic and political interests.

Natural Gas

Natural gas can be burned in a gasoline-powered engine with only minimal adaption. Because it offers more advantages than methanol and enjoys the support of the natural gas industry, the prospects for its widespread use are good. Natural gas supplies are abundant in the United States and many other countries.

Currently, about a half million vehicles worldwide operate on compressed natural gas, known as CNG. Most are in Italy. Countries such as New Zealand, Canada, the United States, and Russia, with large domestic supplies of natural gas, also have a moderate number of retrofitted gasoline-powered vehicles running on this fuel. Virtually all vehicles are bi-fuel, with redundant natural gas and gasoline tanks and lines. The cost of converting a vehicle, including labor, ranges from about $2,000 upward. Retrofitted CNG vehicles are economically attractive only when gasoline is very expensive and the vehicles are used extensively—the advantage of cheap fuel offsetting the expense of conversion.

If natural gas vehicles are to gain widespread acceptance, they must be specifically designed and manufactured for natural gas alone, not burdened by a redundant fuel system. Vehicles optimized for natural gas would have generally lower emissions than gasoline vehicles, about 10 percent greater efficiency because of the fuel's higher octane, and similar power; but they would cost about $700 to $1,000 extra for high-pressure fuel tanks (or cryogenic liquefaction tanks).

Taking into account all the other costs of owning and operating a vehicle, the full life-cycle cost of a factory-made CNG vehicle would average about the same or less than that of a methanol vehicle, and close to that of a gasoline vehicle.[38] Under favorable conditions—high gasoline prices and heavy driving—it would be less expensive than both methanol and gasoline.

On the negative side, CNG vehicles would have a shorter driving range and/or less trunk space than methanol or gasoline vehicles because the CNG tank would have to be larger. Even when compressed or liquefied, natural gas has a much lower volumetric energy density than petroleum.

Environmentally, natural gas should prove superior to both methanol and gasoline. CNG vehicles emit lower levels of greenhouse gases than gasoline vehicles (see Table 2-1). This small advantage is the result of a sharp reduction in carbon dioxide exhaust, which more than offsets the increase in methane emissions from the vehicle and in pipeline leakage.

Natural gas combustion tends to generate much less carbon monoxide compared with gasoline combustion but a similar quantity of nitrogen oxide. It produces a substantial amount of organic gases (analogous to hydrocarbon emissions from gasoline), but these are far less reactive than those from methanol or gasoline combustion and therefore less of an ozone threat.[39]

In principle, it should be easier and cheaper to reduce emissions from CNG engines than from gasoline engines. Indeed, in 1993, a CNG minivan from Chrysler was the first production vehicle to be certified as "ultra-low emission" by the State of California. Once again, however, without strong incentives to do otherwise, automakers are likely to continue their focus on cleaning up the polluting products of gasoline combustion rather than pursuing a new technology.

Start-up barriers to the widespread use of CNG vehicles include the cost of natural gas fueling stations, which are much more expensive than methanol stations—at least $300,000 versus about $30,000. And, as noted above, transitional bi-fuel natural gas vehicles tend to be more expensive than flexible-fuel methanol vehicles.

The future of natural gas vehicles powered by internal combustion engines (ICEs) is uncertain. The benefits are not substantial enough by themselves to inspire the groundswell of governmental support needed to overcome such substantial start-up costs. The success of natural gas vehicles will depend principally on the level of support from the large but sometimes lethargic natural gas industry. It is unlikely that ICE natural gas vehicles will ever dominate the transportation vehicle market.

Alternative Fuel Summary

While ICEs powered by natural gas, methanol, and biomass fuel will not lead to a sustainable transportation system, they could play an important transitional role. They could serve, for instance, as the bridge fuels for powering electric-drive fuel cell vehicles. The creation of a methanol fuel distribution system and natural gas fueling stations could set the stage for fuel cell vehicles powered by methanol and natural gas–derived hydrogen. Also, advances in the technology for storing natural gas in vehicles could be applied to the development of similar storage tanks for hydrogen.

Electric Chaff

Later we will see that electric propulsion technologies are fundamentally more efficient and cleaner than ICEs. But not all electric technologies make sense. Here is a brief review of those that are flawed or unacceptable for widespread application.

Wiring the Roads

Public roadway–powered electric vehicles, that is, those that gain their electricity from wires embedded in the road pavement, are a dead-end technology.[40] While this option might prove economically competitive with others, including conventional gasoline vehicles, it is difficult to imagine how the technology would be deployed in a politically and economically acceptable manner.[41] The most likely home for this technology is self-contained road systems—for example, within large factories, military bases, or airports with many on-site service vehicles. The two principal barriers are strategic: a disproportionate share of the costs are up-front, and strong competitors are emerging that solve the limited-range problem of current battery-powered electric vehicles.

The transition to public roadway–powered vehicles is unlikely ever to get under way because of the huge start-up costs. Thousands of miles of road would need to be retrofitted with large encased cables at over $1 million per lane per mile, and many thousands of dollars spent per vehicle to outfit them to drive these roads. Because automakers and consumers are unlikely to invest thousands of dollars to outfit a vehicle to pick up electricity from nonexistent electrified roads, government would have to retrofit the roads before any vehicles were sold—plus provide large up-front incentives to stimulate the production and sale of these specialized vehicles. It is difficult to imagine any government spending many billions of dollars to rip up roads and disrupt traffic for vehicles not yet off the assembly line.

The large bill could be accommodated if there were good reason to do so. There is not. While roadway-powered vehicles are a possible solution to the problem of recharging batteries, other options now seem more compelling, including improved batteries, fuel cells, and hybridized ICE–electric "powertrains." And we should not underestimate the ability of drivers to accept more limited driving ranges.

Wires Alongside and Overhead

Electricity could also be provided to a vehicle as needed from rails alongside the road or overhead. Indeed, overhead catenary wires have been used for a century to supply electricity to trolley buses and streetcars. These systems could be revived, but in addition to being unsightly, they are too cumbersome for most vehicles. Other more innovative ideas for using powered rails—placing them in the pavement or adjacent to vehicles—present some of the same start-up and aesthetic problems facing systems relying on power from roadways or overhead wires.

— — — — — — — — — — — — —

We have seen that a number of the technologies and strategies receiving the most attention, such as IVHS, alternative fuels, and uniform emission standards, do offer some advantages, and that the advantages would expand if these innovations developed hand in hand—with each other and with various regulatory and pricing strategies. The challenge is to sift the sands and separate what is positive and promising from that which is not. Biomass-based alcohol fuels, for example, merit more government support than corn ethanol, and smart paratransit more support than other smart car technologies. The more important point, though, is that none of the technologies or strategies discussed thus far show anywhere near the promise of electric-drive technologies.

CHAPTER 3

Toward Electric Propulsion

It's no surprise that today's electric vehicles cost more and perform worse than their gasoline counterparts. Gasoline cars have benefited from a century of intensive development; electric cars have been virtually ignored for over seventy-five years. Even today, gasoline cars profit from billions of dollars of research every year while electric vehicles receive a tiny fraction of that.

Why didn't electric motors, despite an auspicious beginning, receive the same attention as gasoline engines? According to Dean Drake of General Motors, at the turn of the century when motor cars were a new invention, electric vehicles outnumbered gasoline-powered vehicles.[1] They were attractive because of the ease with which they could be started and driven. "The electric runabout . . . appears to be the most popular form of automobile for women, at any rate in the National Capital," noted a society columnist in 1904.[2] W. C. Durant, the founder of General Motors, pointed out that gasoline-powered cars were "noisy and smelly, and frighten the horses."[3] But women did not buy many vehicles, and battery technology did not improve nearly as fast as gasoline vehicle technology. By 1910 the heyday of the electric runabout was over. The Ford Model T was now selling for less than half the price of any advertised electric car.[4] By 1915, less than 2 percent of the 2.5 million motor vehicles in operation in the United States were powered by electricity.[5] The electric vehicle industry dwindled away, with the last factory in the United States closing in 1935.[6]

Gasoline-powered cars pulled ahead of the competition as a result of technological innovation. Charles Kettering's electric starter retired the hand crank, engines became more efficient, rubber engine mounts reduced vibration, and advances in carburetion and ignition made gasoline cars easier to drive. Meanwhile, oil discoveries swept away concern about energy supply, and a spreading network of roads rewarded the much greater range of gasoline cars. The only electric vehicles that continued to be produced were forklifts and other equipment that operated indoors at short ranges.

Here matters stood until recently. The invention of semiconductors in the 1950s and continuing improvement in motors and controllers spurred some interest in electric vehicles in the 1960s. General Motors and Ford initiated modest research into electric vehicles, and at the 1966 auto shows General Motors displayed a converted Corvair running on silver-zinc batteries. Meanwhile, a number of small companies and individuals were hand-building electric cars and small trucks, usually by converting gasoline-powered versions.[7] In the mid-1970s Robert Beaumont produced 2,000 electric Citicars. The small car was aggressively promoted. It sold for about $3,000, comparable to the least expensive gasoline cars available in the United States at that time, but it was not well designed or engineered and was severely criticized by *Consumer Reports* and other industry reports. Production ended abruptly after two years.

A notable non-event in EV history was General Motors' announcement of plans to begin commercial production of electric vehicles. It came in 1980, at the height of the oil crisis; the idea was quickly abandoned as oil prices dropped.

Elsewhere in the world, many others were also experimenting with electric vehicles. A small number of electric delivery vehicles were used in Europe and, to a lesser extent, in Japan. British manufacturers, which never stopped producing these vehicles, turned out several thousand electric milk delivery trucks a year. By 1982, an estimated 33,000 were on British roads.[8]

In Japan, the history of electric vehicles has centered more on major automakers. The larger companies all embarked on determined development programs in the 1960s. By the late 1970s, at least four major companies had built advanced prototypes that were reportedly ready for commercial production. They held production plans in abeyance and apparently cut back their research and development efforts during the 1980s, satisfied with building a small number of very small electric vehicles for the domestic market. When California adopted its ZEV mandate in 1990, the major Japanese manufacturers all had substantial experience with electric vehicles. Even so, they were reluctant to market electric vehicles, sharing the concerns of the Big Three that EVs were too costly and risky. Japanese automakers maintained a low profile throughout the EV debates of the early 1990s.

It was in direct response to California's ZEV mandate that major investments in electric vehicle technology were finally made in the 1990s. Even then, the sums were modest by industry standards. Ford and General Motors reported spending a total of $450 million during the first few years of the decade. The total U.S. investment in electric vehicles and batteries by industry and government during those same years was probably less than $1 billion. To provide perspective, consider that the cost of outfitting one major refinery to produce reformulated gasoline would be about the same. Indeed, the U.S. oil industry plans to spend $13.7 billion from 1991 to 2000, and considerably more after 2000, solely to upgrade refineries for reformulated gasoline and diesel fuel and to meet other environmental requirements at refineries.[9]

The Return of the Electric Vehicle

The current surge of interest in electric vehicles is due strictly to one concern—air quality. In 1989, inspired by the realization that alternative fuels were not as clean burning as once thought, Los Angeles City Council member Marvin Braude issued a worldwide bid for 10,000 electric vehicles to be delivered to southern California in the mid-1990s. Two local electric utilities provided initial funding. Braude's request attracted little response from major automobile manufacturers, principally because of their continuing skepticism about the market, and the bid was awarded to Clean Air Transport, an Anglo-Swedish company with a handful of employees. After spending several million dollars to build a hybrid electric-gasoline car, the company was unable to attract the additional funding needed to begin manufacturing. It quietly disappeared in the early 1990s, leaving behind only two working prototypes. By this time, the major automakers had regained interest.

The next key event occurred on January 3, 1990, when Roger Smith, CEO of General Motors, held a press conference to unveil the sporty battery-powered Impact. As he was leaving the conference he reportedly inquired, apprehensively, "You guys aren't going to make us build that car, are you?"[10] The answer, after a subsequent General Motors announcement that it intended to put a derivative of the Impact into mass production in the mid-1990s, was a definitive yes.

The answer came from California air regulators. Encouraged by GM's announcement, the California Air Resources Board (CARB) accelerated its schedule for introducing electric vehicles in the state. In September 1990, after months of public hearings, and relatively passive objections from the auto industry (due in part to preoccupation with debates in Washington over the national Clean Air Act), CARB adopted its rule requiring that a growing percentage of each major automaker's sales in California be ZEVs. ("Major" automakers are those selling over 35,000 vehicles a year in California. The following companies fall within this category, in descending order: General Motors, Ford, Toyota, Chrysler, Honda, Nissan, and Mazda. In 2003, the threshold will drop to 3,000 sales per year, which will affect most European companies and smaller Japanese companies but not specialized companies such as Rolls Royce and Ferrari.) The required percentage was set at 2 percent for 1998, increasing to 5 percent in 2001 and 10 percent in 2003.

In early 1991, shortly after California adopted the mandate, General Motors announced that it had selected an assembly plant in Lansing, Michigan, for production of an electric vehicle based on the Impact prototype. Kenneth Baker, head of the company's electric vehicle program, is reported to have declared at that time, "If you're waiting for us, the waiting is over. We're committed to deliver a product by the mid '90s."[11] The announcement was pivotal; it strengthened the resolve of CARB to stick with the ZEV mandate and signaled to other manufacturers that efforts to delay or suppress the mandate were now doomed. A surge of electric vehicle investment followed. Shortly thereafter, Maine, Maryland, Massachusetts, New Jersey, and New York followed California's lead, adopting the ZEV mandate for themselves.

Resisting the ZEV Mandate

The spreading mandate alarmed the major automakers, including General Motors. They had been resigned to accommodating California, treating it as an expensive experiment. Doubling electric vehicle production to serve those other states was an altogether different matter. With "superbatteries" languishing in laboratories, the spreading mandate meant absorbing thousands of dollars in losses per vehicle on twice as many vehicles. In their eyes, the trend was beginning to look like a nightmare. General Motors, losing tens of billions of dollars, was especially vulnerable.

On December 11, 1992, shortly after ousting CEO Robert Stempel, like his predecessor Roger Smith a strong electric vehicle advocate, General Motors announced that it was delaying production of electric vehicles until the late 1990s. It also announced that it would work more closely with Ford and Chrysler on joint ventures to manufacture electric vehicle components. The joint ventures never materialized. What GM did do was to join other automakers in filing a series of lawsuits against states seeking to adopt California's ZEV mandate. The legal premise for these challenges is a Clean Air Act provision dating back to the 1960s that gives California the right to adopt its own vehicle emission standards, and all other states the right to adopt only California's or the EPA's, but no one else's.[12] Though the latest rulings have gone against automakers, lawsuits are still pending as this book goes to press in late 1994.

Meanwhile, proponents for clean air and electric vehicles, led by Tom

Jorling, head of New York State's environmental agency, tried to out-flank the automakers by turning to a federally created consortium of northeastern states known as the Ozone Transport Commission. With a simple majority vote, this commission had the authority to impose pollution rules across the entire region. The automakers reacted with alarm. They offered a compromise in late 1993. They volunteered to reduce the emissions of gasoline cars sold in those states below what was required if the states abandoned the ZEV mandate. Their compromise proposal was rejected on February 1, 1994. By a nine to four vote, the eleven states, plus Washington, D.C., and its Virginia suburbs, asked the EPA for permission to adopt California's low-emission-vehicle program for their region. A decision from the EPA is expected in late 1994. If permission is granted, each state will have the option of adopting the ZEV mandate for itself.

Meanwhile, the auto industry pursued one other strategy. On September 29, 1993, the Big Three agreed to work together with the federal government to build what became known as the next-generation vehicle. No production schedules or financial or regulatory commitments were made. The expressed intent was to leapfrog vehicles run on batteries and develop those they believed would be more marketable, mainly fuel cell and hybrid vehicles (see pp. 141–143). The unexpressed intent, according to many observers, including the industry trade press, was to undermine both the ZEV mandate and proposed increases in corporate average fuel economy standards.

With General Motors backing off from its Impact program, the other two American automakers adopted a more aggressively hostile attitude toward California's ZEV mandate. In a widely circulated letter to Governor Pete Wilson of California on September 13, 1993, Ford's vice chairman, Alan Gilmour, made the following claims: there are no car batteries "with acceptable cost, range, and life"; the mandate is "an inefficient and cost-ineffective road toward meeting the ultimate goal, which is cleaner air"; governments should not mandate specific technology as a means of achieving public objectives, but should "set objectives and then work with the industry or affected party on finding the most efficient means of achieving those objectives."

This automaker view did not prove compelling in California. After months of often bitter public debate and a raucous two-day public hearing in Los Angeles on May 12–13, 1994, CARB reaffirmed the mandate without changes. Jacqueline Schafer, CARB chairwoman, concluded the hearing:

> We heard from no one who claimed the mandate had not accomplished its stated objective of stimulating technological de-

velopment and innovation. While electric vehicle and battery technology may not have advanced much between the turn of the century and the 1980s, there is no doubt that tremendous advancements have occurred since we adopted the zero emission vehicle regulation in 1990. We heard over and over again that the mandate caused or contributed to these advancements. I don't think we want to take any actions that would slow down or stall this progress . . . [The mandate] must remain in place.

General Motors' chairman, John Smale, conceded that the auto industry lost the battle in large part because of "a lack of credibility on the part of not just General Motors but the other automobile companies. The legislators and regulators seem to feel that they cannot trust the industry to do something."[13]

In retrospect, how does one explain General Motors' flip-flop on electric vehicles from advocate to litigant? Some argue that the company allowed its enthusiasm for retaking market share from the Japanese and for projecting a "green" image to overwhelm its business sense. According to this view, the company underestimated the challenge of designing and building an entirely new propulsion system. Indeed, even after the Impact was completed, GM found it necessary to re-engineer the entire prototype vehicle, investing much more time and money in converting it to a production vehicle than had been anticipated.

Toward 1998

As of now, late 1994, the production plans of the "Big Seven" automakers (those required to meet the ZEV mandate) remain unknown, with one exception: Chrysler announced in May 1994 that it would build an electric version of its minivan. It is expected that all seven, as well as many other automakers, will build variations of the electric vehicles they have already built and displayed as prototypes. For instance, Ford will probably produce some version of its microvan, the Ecostar, and General Motors some version of its Impact sports car. Between 1998 and 2000, the major companies are likely to build at least one subcompact or compact car, plus one small van or pickup truck.

Not all electric vehicles produced for 1998 will come from the Big Seven. Some will be supplied by smaller automakers, such as Peugeot, Mercedes, Mitsubishi, and Volvo, which are not affected by the 1998 mandate. Start-up companies and companies without automotive experience may also enter the industry; they may sell converted gasoline cars, electrified "gliders," and mini-cars or neighborhood electric vehicles.

Electric vehicle prototypes from major automakers (*clockwise from top left*): Chrysler TEVan, Nissan FEV, General Motors Impact, Honda EVX, Ford Ecostar, and Toyota EV-50.

In the early 1990s, small start-up companies were converting an increasing number of gasoline cars to electricity, but total sales amounted to less than 1,000 by 1994. Most conversions were sold to electric utilities and government agencies that wanted to test and promote them locally. Small conversion companies will need strong technical capabilities if they are to survive. Solectria, a successful pioneer company in Massachusetts headed by young James Worden, has built state-of-the-art vehicles that have won virtually every electric vehicle race in the United States since the early 1990s. But even Solectria will have to focus on designing and selling specialized electric vehicle components, or serve as a dealer for a larger company if it is to succeed over the long term.

Other small companies will compete by purchasing "glider" vehicles, those without engines and powertrains, from major automakers. They will install batteries and an electric driveline into these gliders for sale under their own name or the glider manufacturer's. U.S. Electricar, a small company in northern California, reportedly signed an agreement with General Motors in mid-1994 to do just this.

Another type of company likely to enter the ZEV industry is the manufacturer of very small cars and trucks. As indicated earlier, mini-vehicles are much less expensive to build than larger vehicles and are well suited to the low energy density of batteries. In the early 1990s a number of European companies began selling "neighborhood" mini-vehicles in the United States. Several U.S. companies have built prototypes and plan to enter the business as well.

Of all the major automakers not bound by the 1998 ZEV mandate, Peugeot-Citroen has been the only one to make a firm commitment to electric vehicle production.[14] After selling about 1,000 small electric vans in Europe between 1989 and 1993, Peugeot-Citroen announced in March 1994 that it intended to start selling a subcompact electric vehicle in December 1994, and to ratchet up production to 10,000 vehicles per year soon thereafter. Some of these vehicles are intended for California and other locales in the United States. The first vehicles will essentially be converted gasoline cars. The plan is to redesign them gradually so that by the year 2000, when annual production runs reach 50,000, they will be optimized for electric drive.

An important lure for all these companies is the ability to sell ZEV credits. For each vehicle sold, CARB provides a company with a ZEV credit. Credits can be sold to any of the Big 7 manufacturers. And because the fine for not selling a ZEV is $5,000, and the loss per vehicle is expected to be greater than $5,000 per vehicle initially, the Big Seven should be willing to pay up to $5,000 for each vehicle credit. The trading of ZEV credits could be attractive to all concerned. It allows a company such as General Motors to meet part (or all) of its ZEV responsibilities without paying an embarrassing fine and without incurring large losses. And it provides small, cash-strapped electric vehicle companies with a substantial cash subsidy for every vehicle sold, even small neighborhood electric cars.

Environmental Benefits

As indicated earlier, the primary premise for government support of electric vehicles has been air quality. Indeed, the sole legal justification for the ZEV mandate is controlling air pollution. Electric vehicles will not disappoint on this count. But air quality alone, as Ford vice chairman Gilmour asserts, is probably not sufficient justification to mandate electric vehicles. With emissions from gasoline vehicles continuing to decline and air pollution receding in most urban areas, it is difficult to justify electric vehicles solely as an air pollution control strategy, except in Los Angeles and a few other metropolitan areas.

Electric vehicles do, however, offer other strong benefits that are ignored by the marketplace. One is the dramatic reduction in oil consumption and petroleum imports that their widespread use would bring about. Much less oil would be needed because only a tiny proportion of electricity is generated from oil—less than 5 percent in the United States. The other major nonmarket benefit would be lower greenhouse gas emissions.

Air Quality

Automobiles and light trucks are responsible for about half of all urban air pollution. Today's vehicles, powered by gasoline and diesel fuel, emit the vast majority of human-produced carbon monoxide, about half the hydrocarbon and nitrogen oxide pollutants, and a small proportion of particulate matter and sulfur oxide. Switching to electric vehicles would dramatically reduce urban air pollution, even if coal-fired power plants were to generate much of the electricity.

Regardless of the type of power plant, fuel, and emission controls, battery-powered electric vehicles would practically eliminate emissions of carbon monoxide and hydrocarbon (also referred to as reactive organic gases and volatile organic compounds) and would greatly diminish nitrogen oxide emissions. While electric vehicles could contribute additional sulfur oxide and particulate matter to the air in some cases, this would not be too serious because cars are responsible for only about 1 percent of these air pollutants. A large decrease in hydrocarbon, carbon monoxide, and nitrogen oxide emissions from light-duty highway vehicles would in general have a much greater impact on urban air quality than would increases in vehicle-related sulfur oxide.

As indicated in Table 3-1, the air quality benefits of electric vehicles would vary depending on circumstances. For instance, sulfur oxide emissions would be minimized if natural gas were used to generate electricity (as in Japan), but would increase if coal steam plants served as the principal source (as in the United Kingdom and United States).

If electricity were produced from solar, nuclear, wind, or hydroelectric power, it would be essentially nonpolluting to the air. If electricity were generated by a combination of fossil and nonfossil energy, the pollution reduction would still be substantial.[15]

TABLE 3-1
Percentage Change in Emissions from Gasoline-Powered Vehicles to
Battery-Powered Electric Vehicles

	Hydrocarbons	Carbon Monoxide	Nitrogen Oxides	Sulfur Oxides	Particulates
France	−99	−99	−91	−58	−59
Germany	−98	−99	−66	+96	−96
Japan	−99	−99	−66	−40	+10
United Kingdom	−98	−99	−34	+407	+165
United States	−96	−99	−67	+203	+122

Sources: Choosing an Alternative Fuel: Air Pollution and Greenhouse Gas Impacts (Paris: OECD, 1993). U.S. estimates are from Q. Wang, M. DeLuchi, and D. Sperling, "Emission Impacts of Electric Vehicles," *Journal of the Air and Waste Management Association* 40: 1275–84(1990).

Notes: Analysis is for cars. Emissions from the full fuel cycle are accounted for, including tailpipe, evaporative, and refinery emissions associated with gasoline use in cars.

Not all studies of electric vehicles have come to the same conclusion.[16] Those with less positive results base their calculations on today's primitive converted vehicles, which are energy guzzlers compared with factory-made vehicles of the future; power plants that emit pollution at 1990 levels, rather than the substantially lower levels required by current and future emission rules; and/or a high proportion of coal-fired power plants. But these are short-term phenomena.

In some locales, the air quality impact of electric vehicles can be extraordinarily positive. Los Angeles is one example. CARB estimates that electric vehicles in Los Angeles would come close to eliminating not only carbon monoxide and hydrocarbon but also nitrogen and sulfur oxide emissions.[17] Even particulate emissions would be greatly reduced. Why? First, about 80 percent of the electricity used during daytime in Los Angeles and about 33 percent used at night comes from outside the region. Second, most of the local electricity is produced by natural gas, less polluting than any other fossil fuel. And third, power plant emissions in Los Angeles are tightly controlled. One might argue that Los Angeles is simply exporting much of its air pollution, and that is true. But the damage from one unit of pollution elsewhere is far less than the damage from one unit of pollution in the fragile atmosphere around Los Angeles.

However positive the impact of electric vehicles on emissions may be, their impact on air quality is far greater. That is because many power plants are located far from populated areas, and a large proportion of emissions from electricity production are released at night when sunlight is not present to form ozone and when people are indoors and not exposed. ICE vehicle emissions are "in our face," while EV emissions tend to be far off and at night. Widespread use of battery-powered electric vehicles would be an extraordinarily effective air quality control strategy almost everywhere.

Greenhouse Gas Emissions

World leaders seem to be moving toward a consensus that only through major reductions in greenhouse gas emissions will civilization avoid economic and ecological disaster. In early 1993, President Clinton endorsed an agreement from the 1992 Rio de Janeiro Earth Summit, proclaiming his determination to roll back U.S. greenhouse gas emissions to 1990 levels by 2000, and still further thereafter. As of 1994, meeting this goal by the year 2000 would require a reduction of emissions by 7 percent, a formidable task given the current trend of 1 to 2 percent annual *increases* in greenhouse gas emissions.

Battery-powered electric vehicles would provide modest greenhouse gas savings if introduced today. As the energy efficiency of the vehicles and the power plants continued to improve, the savings would be more substantial. More importantly, the introduction of electric drive into vehicles would create the opportunity for dramatic savings not possible with any other transportation or energy option.

Power plants that burn fossil fuels emit the most greenhouse gases (see Table 3-2). On a per-mile basis, electric vehicles using coal-fired power plants would increase greenhouse gas emissions slightly. Electric vehicles using natural gas would lower greenhouse gas emissions relative to conventional gasoline vehicles, mainly because of the low carbon-to-hydrogen ratio of natural gas. Electric vehicles powered by nuclear or hydroelectricity would have nearly zero greenhouse gas emissions.[18] If powered by a *marginal* mix of electricity sources (projected to be in use in the United States by 2000), electric vehicles would be somewhat superior to gasoline vehicles.[19]

Technology: Current Status and Prospects

It is widely accepted that internal combustion engines will eventually be supplanted by electric drive. The only question is when. Electric drive is inevitable because it is much more efficient and has many fewer moving parts. The fact that is it quiet and without emissions serves only to hasten the transition.

The energy efficiency advantage is due to several factors. First, electric motors are over 90 percent efficient, while ICEs are 15 to 25 percent effi-

TABLE 3-2

Change in Greenhouse-Gas Emissions from Gasoline-Powered to Electric Vehicles

Electric Vehicle Fuel/Feedstock	Change (percent)
Solar and nuclear	−90 to −80
Natural-gas plants	−50 to −25
Current U.S. power mix	−20 to 0
New coal plant	0 to +10

Source: Mark A. DeLuchi, "Emissions of Greenhouse Gases from the Use of Transportation Fuels and Electricity" (Argonne National Laboratory, Center for Transportation Research, Argonne, Illinois, 1991, ANL/ESD/TM-22).

Notes: Emissions from vehicle and materials manufacturing are assumed to come from fossil fuels. If these manufacturing processes were excluded, or were assumed to use nonfossil energy such as solar, then electric vehicles would provide an additional 10 percent reduction (see Table 2-1 for greenhouse-gas emissions from other transportation energy options).

cient. Second, about 10 percent of the energy used by ICE vehicles is during idling; electric vehicles use no energy during idling because the motor shuts off. Third, unlike ICE vehicles, electric vehicles can recapture heat energy lost during braking. About one-third of the energy used by a motor vehicle is lost during braking; electric vehicles can recover up to half that energy with regenerative braking. And fourth, the absence of a transmission in electric vehicles saves another 6 percent of energy consumed in ICE vehicles.

Partially offsetting the efficiency gains of electric vehicles are the low efficiencies at power plants. Oil refineries are about 90 percent efficient, compared to efficiencies of about 33 percent with today's power plants fired with oil, natural gas, and coal. But oil refineries are not expected to become more efficient, while new power plants are already exceeding 40 percent efficiency and will likely top 50 percent soon.

Without major fanfare, electric drive has become cheaper and far more efficient. For example, advances in power electronics have made possible alternating-current (ac) drives that are cheaper, easier to maintain, and more compact, reliable, efficient, and adaptable to regenerative braking than the direct-current (dc) systems used in virtually all electric vehicles in the early 1990s. By one account, the weight, volume, and cost of the electric propulsion motor and associated electronic controller were reduced by an estimated 60 percent in the 1980s.[20] The electric motor-controller combination is now smaller and lighter than a comparable internal combustion engine, as well as being cheaper to manufacture and maintain.

Electric vehicle technology is accelerating in response to the California ZEV mandate. As of 1994, all major automakers in Europe, North America, and Japan had major research and development programs in this area.

Consider General Motors' Impact, discussed earlier. With an ac motor, a compact, efficient inverter, and an ultra-high-efficiency design, it achieves a seventy-mile urban range, even using lead-acid batteries, performing better than a comparable gasoline vehicle. An early prototype accelerated to sixty mph in eight seconds, faster than Nissan's 300ZX sports car. *Popular Science* described it as "possibly the best-handling and best-performing small car that GM has ever turned out."[21]

There is room for further improvement in the motor, power electronics, transmission, aerodynamics, tire friction, and auxiliaries of the electric vehicle, as well as in the battery. The greatest potential for improvement lies with batteries and auxiliary equipment, especially for heating and cooling the vehicle. Heating and cooling systems are not a problem in the internal-combustion vehicle because they can draw on waste heat and surplus power from the engine. They are a problem in the electric vehicle, whose motor does not generate enough heat for warming the cabin and whose battery does not store enough energy for conventional cooling and heating systems. The type of air conditioner and heater used in today's gasoline vehicle could reduce the range of an electric vehicle by as much as 20 percent or more. There are creative solutions being investigated to minimize the need for battery-supplied electricity. These include a high-efficiency heat pump, window glazing to reduce solar heat inside the cabin, air cooling through the seat instead of throughout the entire cabin, and dissipation of heat buildup in the parked vehicle using a solar-fueled fan.

Batteries: The Weak Link

At the present time the most expensive component in the electric driveline is the battery. Currently available lead-acid batteries are not satisfactory in cars because they last less than two years and 20,000 miles and do not store enough energy to power a full-size vehicle very far or fast. But major improvements are expected. In the past decade a number of new battery types have been commercialized for use in portable computers, cameras, camcorders, and cellular phones. The dramatic improvements in these small batteries suggest that major improvements are also possible with the large batteries suitable to electric vehicles.

In the United States, the most impressive and significant research in this area is being done by the Advanced Battery Consortium (U.S. ABC), launched in 1991. U.S. ABC's goal is to increase energy and power capability, extend the life, and reduce the cost of batteries as they are scaled up to sizes suitable to electric vehicles. The budget calls for $230 million to be spent by 1995 and $100 million more per year into the early years of the next century. Funding is provided by the federal government (50 percent), the Big Three automakers (23 percent), battery companies (20 percent), and electric utilities (7 percent).

Other battery consortia were formed shortly after U.S. ABC. The Japanese Ministry of International Trade and Industry (MITI) started a ten-year research program, with about half the funding of the American one, while smaller efforts were launched in Europe and by the non-aligned Advanced Lead-Acid Battery Consortium headquartered in the United States.[22]

Battery technology for electric vehicles is inadequate largely because of lack of effort. Until U.S. ABC was formed, virtually all battery research focused on small batteries. The explanation is simple: the market for small batteries to power a rich array of consumer products was tens of billions of dollars, while the market for traction batteries to power golf carts and forklifts approximated 1 percent of that, about $200 million. Until California's ZEV mandate, the electric vehicle market was not lucrative enough to justify much attention.

Less than $10 million a year was being spent on battery research aimed specifically at the vehicle market, and most of that went to improving the primitive techniques employed in building battery packs for golf carts and slow-speed industrial vehicles.[23] Typically, these packs were conventional six-volt "engine-starter-type" lead-acid batteries wired together in trays or boxes and without integrated controls. The battery packs put together for the G-van, the only widely available electric vehicle prototype of the late 1980s, were of such poor quality that they were practically being given away by 1993.

Electric vehicles require batteries with much greater energy density and sophistication of design than those that power golf carts and forklifts. Nonetheless, the future of battery development remains uncertain, in part because commercial rivalry has resulted in secrecy and politicization of battery research. There is also uncertainty regarding the relative importance of different battery attributes—should batteries be developed for low cost, for high energy density so that range can be extended, or for high power density for use in hybrid vehicles? The weak commitment of the auto industry, as indicated by their willingness to provide only one-fifth of U.S. ABC's funding, does not help matters.

In the near term, at least through 2000 and probably beyond, lead-acid batteries are likely to dominate the electric vehicle battery market because of their relatively low cost and high reliability. Continued improvements are expected, with research partly funded by the Advanced Lead-Acid Battery Consortium. This group, which has a diverse membership from around the world, is only spending about $5 million a year.

There is no consensus as to which type of battery will ultimately prove superior for electric vehicles (see Table 3-3). The list of candidates is long. It includes batteries with solid, liquid, and gas electrolytes, high and moderate temperatures, replaceable metals, and replaceable liquids, as well as batteries made with a wide variety of materials. At least twenty distinct battery types have been suggested as candidates for electric vehicles. But what looks promising in a small cell often fizzles when scaled up for a vehicle. The reality is that the underlying science of battery technology is, in the words of one *Business Week* reporter, "fiendishly tricky." Battery breakthrough announcements are so common there ought to be a battery-of-the-month club.

Batteries that have received considerable attention but are not among the most promising include nickel cadmium, nickel iron, and sodium sulfur. Of these, the sodium-sulfur battery has been the greatest disappointment. Two large European companies invested an estimated $500 million in development of this battery over three decades, and it was considered a leading contender as recently as 1992.[24] One set of problems is related to its high operating temperature of 300–350° centigrade. If the battery is left unused several days, it loses a considerable amount of energy; if left longer, permanent damage is possible. The high temperature is also hazardous. In 1994, out of 34 sodium-sulfur batteries installed in Ford Ecostar prototypes, two caught fire within months of being put in use. Another stubborn problem is that the metal-ceramic seals for sodium-sulfur battery cells corrode after two years. Ford, which invented this technology in the 1960s and then abandoned it to the Europeans, still maintains hope that the technical and safety problems of sodium-sulfur batteries can be solved, for it is potentially high performance and low cost.

A variation of sodium-sulfur technology being tested in vehicles in Europe in 1994 that appears somewhat more promising is sodium nickel chloride. Like sodium sulfur it operates at high temperatures (250–320° centigrade), but it is more resilient and appears not to be subject to corro-

TABLE 3-3
Goals for Advanced Batteries Compared to Today's Lead-Acid Batteries

	Today's Lead Acid	ALABC Goals	U.S. ABC Mid-Term Goals	U.S. ABC Long-Term Goals
Energy density (wh/kg)	40	50	80	200
Power density (w/kg)	100	150	150	400
Cycle life (80% depth of discharge)	300	500	600	1000
Recharge time (hours)	6–8	4[a]	6	3–6
Life (years)	2	3	5	10
Price ($/kwh)	120[b]	150	150	100
Battery cost ($/kwh/yr)[c]	60	50	30	10

[a] Four hours for 100 percent; 15 minutes for 80 percent; 5 minutes for 50 percent.

[b] Lead-acid batteries can be purchased for less, but they are not well suited to the deep-discharge patterns of electric vehicles; when used in electric vehicles, such batteries have very short lives.

[c] Battery cost per year is not an ALABC or U.S. ABC goal; it is calculated by dividing battery price goals (in $/kwh) by life (in years).

sion. Unlike the sodium-sulfur battery, it can be allowed to cool with no harmful effects to the battery. Sodium-nickel-chloride batteries have relatively high energy and power density—comparable to mid-term U.S. ABC goals—and may prove economical when mass-produced.

Other possible battery contenders include zinc bromide and zinc air, but they have not been thoroughly tested or developed. The zinc-bromide battery, which has high energy density but low power density, would be most economical if hybridized with a device to provide power surges for hard accelerations and hill climbing. The battery materials are inexpensive. At least one European and one U.S. company are working on this battery. It has been used in only a few vehicles.

Zinc-air batteries have received equally scant attention. They are recharged by replacing the zinc electrodes; the zinc is then recycled for future use. Replacement, though relatively simple, would probably have to be done at zinc retail outlets, not at home. The battery has a high energy but low power density, and it is expensive and bulky. Zinc-air batteries are being developed in Israel and the United States, and tested in Germany by the postal service.

Leading contenders for the twenty-first century appear to be nickel-metal-hydride and lithium-based batteries; because of their potential for major improvement, they have received most of the U.S. ABC funding, as well as substantial grants and contracts from other U.S. government agencies and Japanese and European battery consortia.

Nickel-metal-hydride and lithium polymer batteries were developed originally for consumer products such as cameras, camcorders, and other small consumer electronic products; in the early 1990s they largely supplanted nickel-cadmium batteries.

Of this new generation, nickel-metal-hydride batteries are likely to be the first made commercially available. Nickel metal hydride has the long life of nickel cadmium without its expense and toxicity. Nickel-metal-hydride batteries were being tested in vehicles in 1994, and that same year they were used in a Solectria car that won a major electric vehicle race competition. General Motors and Honda have both signed contracts to purchase nickel-metal-hydride batteries from Ovonics, a Michigan company; Honda reportedly has a similar agreement with a Japanese company; and a German company is also developing the technology. Ovonics announced in early 1994 that its batteries would be commercially ready in 1998 at the mid-term price goal set by U.S. ABC, though many experts remain skeptical that the cost can be reduced so low so soon.

Lithium emerged as a promising material for rechargeable batteries in the 1970s. Lithium batteries have the potential to be inexpensive and high performing, though major technical problems such as corrosion, safety, and scaling up remain unsolved. High-temperature lithium-metal-sulfide batteries, which promise to be relatively cheap and to perform better than sodium-sulfur batteries, are being developed by at least two major companies, SAFT and Westinghouse. Lithium-polymer technology received about half the funds awarded during U.S. ABC's first few years. It is being pursued by a number of companies in Europe, Japan, and the United States.

Japan's MITI, with half as much funding as U.S. ABC, is reportedly concentrating on lithium batteries, while another Japanese consortium, including Toyota, has its eyes on nickel metal hydride. The European battery consortium is targeting a mix of sodium-, nickel-, and lithium-based batteries. According to one American battery expert, Japan has taken the lead in the race to find the automotive battery of the future,[25] but there is no consensus as to whether this indeed is the case.

The future of advanced batteries is worrisome not only because of limited R&D investments but also because of R&D strategies. The central goal of the U.S. ABC and MITI, for instance, is to develop batteries that will allow electric cars to compete head-to-head with gasoline cars. The focus is on batteries with high energy density and long life. Is this the best strategy? Not necessarily. Perhaps the primary emphasis should be low cost. If it is true, as it appears to be, that consumers are willing to accept

A French electric vehicle recharging station.

short driving ranges, then a cheap battery with low energy density would sell more cars than an expensive battery with high energy density.

Another issue has to do with power. The battery consortia are not pursuing batteries with very high power density. The reason, again, is the overriding desire of automakers to attain long driving ranges, which requires high *energy* density. But vehicles with hybridized ICE–electric powertrains need high power, not high energy—that is, batteries that can provide an occasional surge of power for sharp accelerations and hill climbing. As indicated in Chapter 6, the peculiarities of government regulation already stack the odds against ICE-hybrid vehicles. Ignoring power-dense batteries in favor of those with high energy density further diminishes the prospects for hybrid vehicles.

And lastly, what about batteries for very small cars and trucks? What batteries would be best suited to their use? That question has not even been asked by the battery consortia.

No matter how successful these various efforts are, the battery will always be inferior to gasoline as a medium for energy storage. The most advanced batteries of the future will probably never exceed 4 percent of the energy density of gasoline. This does not mean that electric vehicles are inherently inferior, though.

First, the 4 percent figure overstates the difference, because electric vehicles are likely to be at least four times as energy efficient as comparable ICE vehicles. Thus future batteries will be closer to 16 percent of the energy density of gasoline. Second, electric vehicles don't have to be comparable. Households could, for instance, own a short-range city car, a long-range minivan, and perhaps a small neighborhood car. Vehicles with shorter driving ranges and smaller mass require only a fraction of the energy of current gasoline-powered vehicles. Third, batteries can be used for peaking if they are combined with very small ICEs that run nonstop or with zero-emission fuel cells. And fourth, the energy consumption of vehicles can be reduced considerably below current levels by improving the efficiency of accessories, tires, aerodynamics, and so on.

A number of strategies related to battery recharging might also be pursued to offset the low energy storage of batteries, including the use of mechanically rechargeable batteries, swapping discharged batteries for fully charged ones, and developing rapid-recharge technology. These techniques hold varying degrees of promise, though none is likely to be a cure-all.

Mechanically recharged batteries include the metal-air type, such as the zinc air mentioned earlier. This type is analogous to the gasoline tank—after each discharge the battery's metal anode is simply replaced in an operation that could take only a few minutes. Metal-air batteries show some promise but have major drawbacks, not least the need for an extensive metal storage and recycling infrastructure.

Batteries could be swapped at stations equipped to extract and replace the packs quickly. The station, automaker, local electric utility, or some other company could own and lease batteries to the vehicle owner, with an added fee charged for each swap. Although this sort of arrangement has caught the public's eye—one-third of all the entries in a national electric vehicle design competition in April 1993 featured battery swapping—it is the least attractive strategy for responding to limited driving ranges. One potential problem is getting battery suppliers to agree on a standardized battery pack, especially during a transition period when rapid improvements are being made in batteries and vehicles. Another is devising a formula for battery-exchange user fees, given the difficulty of calculating the remaining life of a battery and any damage caused by the user.

Fast recharging is likely to play a more prominent role. Recent research has resulted in the development of sophisticated "smart" charging systems that are fast, do little damage to batteries, can be used in many different types of batteries, and even extend battery life. Until the early 1990s, it was thought that rapid recharge would require huge surges of current that might damage batteries, especially if they were recharged more than a few times. Since then, "pulse" techniques have been developed that continually monitor the battery as it is being charged to prevent overcharging and damage. Bursts of energy are applied sequentially in a manner that allows heat to dissipate during the rest period. It is probable that a 50 percent battery recharge could be completed in five to ten minutes without shortening the life of the battery, and possibly even extending it. However, the cost of the recharge equipment and the high power requirements would probably preclude home use.

Fast recharging and battery swapping might be attractive to some fleets (for instance, taxis and buses) and in some other circumstances. Fast charging could become a fixture at fuel stops and rest areas along major intercity highways. The problem is the high cost of providing battery exchanges and fast recharging; it would be especially high if spread over just a few patrons, a likely scenario in the early years of a transition. Because the cost would be higher than wall-plug recharging, and because, as indicated below, most households would use a gasoline vehicle for long trips, where would the demand come from? No doubt some fleets will turn to fast recharging and battery swapping, and no doubt rapid recharge stations will be psychologically comforting curiosities subsidized during the early stages of electric vehicles, but these will not be the primary responses to current battery limitations.

In summary, battery shortcomings can be overcome in a variety of ways—by making vehicles more energy efficient, lowering expectations about driving range and performance, and adopting smarter, faster, and quicker recharge techniques.

The Cost of Electric Vehicles

The essential question is this: Will cost reductions resulting from improved technology and mass production bring the cost of battery-powered electric vehicles down to that of gasoline vehicles? The answer is a definitive no—if we are talking about the initial purchase price. However, if we consider life-cycle costs—all costs calculated on a per-mile basis—the answer is a qualified yes.

What can be said with confidence is the following: electric vehicles will have much lower operating costs and a longer vehicle life than gasoline cars, and, minus batteries, will be less expensive. Chrysler's vice president for electric vehicles, reflecting on short-term costs, notes that "electric minivans without a battery will cost the company as much as gasoline minivans. . . . The cost of the electric motor and controller will about match the cost of the gasoline engine, transmission, fuel and exhaust systems being removed."[26] Ford's director of electric vehicle planning and program office, projecting a little further into the future, comments, "It's not hard to see that we can build an electric vehicle that's cheap, or maybe even slightly cheaper, than our current internal combustion engine vehicles. . . . The bottom line on the cost is in the battery."[27]

Batteries are expensive. As the Ford executive implies, the great uncertainty in projecting the price of battery-powered electric vehicles is the battery itself. Will batteries be so expensive as to offset other cost advantages? The answer depends on a variety of factors, including how much a vehicle is driven, its life and size, its driving range and acceleration capabilities, as well as how far and fast the cost of batteries drops. In general, large electric cars and trucks will always be more expensive than their ICE counterparts (ignoring for now the environmental and energy benefits), principally because they require so much energy and therefore so many batteries. However, considering total life cycle, smaller electric vehicles could well be cost competitive with comparably sized gasoline vehicles early in the next century, if production is scaled up by that time.

Life cycle is a more economically accurate basis for comparing costs, though it would have been misleading, until recently, to emphasize life cycle. What has changed is how people buy cars. Whereas in previous times people purchased vehicles, the trend is now toward leasing. By 1994, some 25 percent of new cars purchased in the United States were leased. The percentage was even higher for fleets and much higher for expensive cars, exceeding 70 percent for some models. Leasing is widely expected to become even more common, according to the automotive trade press. This trend is critical to the future of electric vehicles. Leasing is a convenient mechanism for spreading out the high cost of batteries.

Table 3-4 gives us an idea of how costs differ for comparable gasoline and electric vehicles.[28] The comparison is designed to be as fair as possible. For instance, it was assumed that both vehicle types are driven over the same standard driving cycle. This does not necessarily mean that acceleration capabilities are equal, however. In theory, electric vehicles can be designed to have the same acceleration as gasoline vehicles, but doing so requires sacrificing driving range. Therefore, in practice and in these calculations, electric vehicles will generally have somewhat lower maximum power, even though they can follow the same driving cycle.

In Table 3-4, the near- and mid-term electric vehicles have 10 percent longer life than the gasoline Ford Escort (measured in miles). The long-term electric vehicle has 15 percent longer life. The electric vehicles are all assumed to have 70 to 85 percent of the high-end acceleration of the gasoline car. That is, they have less power at freeway speeds, though they

TABLE 3-4
Costs of Subcompact Electric Cars versus Gasoline-Powered Cars

	Gasoline	Near Term		Medium Term		Long Term	
		Low	High	Low	High	Low	High
Battery	—	Lead acid		Nickel-metal hydride		Lithium sulfide	
Range (miles)	300	100	100	75	125	100	200
Purchase cost							
with battery ($)	12,944	17,761	21,775	15,769	21,124	15,685	18,916
Cents/mile	26.7	29.9	36.8	25.8	33.0	27.8	29.9

Note: The three scenarios incorporate and are premised upon near-term advanced lead-acid batteries, and the mid-term and long-term battery goals established by the U.S. ABC.

probably would have better acceleration at lower speeds due to the unique power characteristics of battery-drive systems. The electricity cost in all scenarios is six cents per kilowatt hour.

The cost results in Table 3-4 should be viewed as scenarios, neither the most optimistic nor the most pessimistic; they are subject to guesswork because of uncertainty with regard to technological and manufacturing progress as well as how the vehicles would be used. For example, costs would be lower if the lifetime of the electric vehicle were to exceed that of the gasoline vehicle by more than the 10 and 15 percent assumed here (many electric fleet vehicles last two to three times longer than their gasoline counterparts), or if the cost of batteries were lower than assumed, or if the ratio of the retail price to the manufacturing cost were lower.

The efficiency of the drive-line and the nature of the driving cycle are also important. Because electric vehicles tend to be more energy efficient than gasoline cars at speeds under 20 mph and less efficient at speeds over 50 mph, electric vehicles are more attractive for congested traffic and neighborhood travel but less so for high-speed freeway travel. This difference is important, because vehicles targeted for lower-speed applications could have smaller, cheaper, and lighter batteries and structural supports.

In summary, electric vehicles could become cost competitive with gasoline vehicles in the foreseeable future.[29] An analysis embracing all external environmental costs would likely indicate that battery-powered electric vehicles would be preferable to gasoline vehicles in many situations in most metropolitan areas in the near future.

Where Would the Electricity Come From?

More electric vehicles does not necessarily mean more power plants. Batteries can be recharged during off-peak hours when excess generating capacity is available. Indeed, the prospect of recharging at night presents electric utilities with an opportunity to operate more efficiently.

Recharging at night could be reinforced by charging drivers much lower rates at that time. This is already standard practice with industrial and commercial electricity customers; many utilities have instituted lower off-peak rates for residential users as well. There are several ways to institute this system, for example, by installing "smart" chargers in the home that automatically turn on during off-peak times or at times signalled by the utilities. Smart chargers are attractive to utilities because they can be used to even out demand through the entire night.

A study conducted by Professor Andrew Ford for Los Angeles–based Southern California Edison, the country's second largest electric utility, found that the company could accommodate at least 1 million electric vehicles in 2010 without adding electricity capacity to its existing resource plan.[30] One million vehicles would represent one-sixth of vehicles projected to be operating in the service area at that time. He found that up to 2 million vehicles could be accommodated if smart chargers were used.[31] Ford found that adding 2 million smart-charged electric vehicles to the system by 2010 would cause the average electricity rate to drop by 1 to 3 percent, a direct result of more efficient use of power plants.

These findings may be somewhat optimistic because Ford ignored the impact of electric vehicles on the distribution system. Also, his study assumed the use of advanced batteries that require less recharging and relatively little charging during the day. In practice, many people would recharge during the day, even with very high rates, for the simple reason that electricity is such a small part of the total electric vehicle cost and because they will occasionally want extra range during the daytime when they travel more heavily. Even so, a conservative estimate is that at least 10 percent of the vehicle population in a region could operate on electricity without creating the need for more generating capacity. Electric vehicles offer an opportunity for more efficient utility systems. Electric companies should be—and are—intrigued by this prospect.

Is There a Market for Electric Vehicles?

In the early 1990s the Ford Motor Company was predicting that only 1 percent of the American population would buy electric vehicles.[32] It arrived at this conclusion by way of focus-group interviews and conventional surveys. More sophisticated studies using elaborate arrays of hypothetical questions come to a similar conclusion: that very few people would buy a vehicle with limited range.[33] These findings are misleading.

Virtually all of these studies have serious methodological flaws. They assume that the electric vehicle will substitute directly for the conventional gasoline vehicle, ignoring the fact that most households have two or more vehicles available that can be used for different purposes. They also fail to explore how preferences and attitudes will change as people gain more information about, and experience with, electric vehicles.

Focus groups and conventional mail and telephone surveys provide insights into the attitudes and preferences of consumers with respect to minor product changes, such as car shape or accessories, but they are not useful in investigating the demand for unfamiliar products. Most people have never driven an electric vehicle, much less thought about such phenomena as home recharging. Responses are poorly informed and, like all opinions and preferences that are not strongly held, subject to change. Even the best of these studies arrive at nonsensical answers. Some, for example, found that buyers would have to be compensated about $15,000 before they would purchase a vehicle with limited range.[34] This finding suggests that a gasoline car costing more than $15,000 but with a small gasoline tank allowing only a short range could not be given away. In short, conventional market studies are highly unreliable and virtually useless for longer-term prediction.

In many ways, the introduction of the electric vehicle to the car market is analogous to the introduction of the microwave oven to the oven market, for both are appliances with an entirely new set of functional features. Microwave ovens took off slowly but steadily, consumer sales increasing from near zero in the early 1970s to three-fourths of American households by 1992. Initially, consumers were concerned that microwave ovens would not give food the taste that conventional ovens imparted. To win consumers over, special cooking classes were given to demonstrate microwave technology. People discovered that the comparison with heat ovens was misguided. The microwave oven couldn't roast a chicken to a golden crisp but it could steam vegetables, defrost, and reheat dishes beautifully. In other words, it didn't displace the traditional oven, it complemented it. The microwave became an additional oven, reallocating old tasks and resulting in entirely new ways of cooking. Like the initial microwave-oven market research, much electric vehicle market research has failed to identify a market for this "radical" new technology. That's because of a flaw in the market research, not because a market doesn't exist.

Research done at the University of California, Davis, paints a more optimistic picture. It suggests that once people are exposed to electric vehicles and to accurate information about them, they will recognize their advantages and buy them. In a test-drive clinic held at the Rose

Bowl in June 1991, drivers were pleasantly surprised by the quality and performance of the electric vehicles they drove—and those were fairly primitive converted cars, a Geo Metro and a Ford Fiesta. Sixty-one percent said their opinion of electric vehicles improved after the test drive, while only 16 percent said their opinion worsened.[35]

In the early 1990s my colleagues and I, with the assistance of Professor Martin Lee-Gosselin of Laval University in Canada, developed a research technique to investigate how people would actually make decisions about buying and using electric vehicles, especially in the 60 percent of U.S. households that own two or more cars. Our objective was to resolve the apparent paradox between surveys that show most motorists travel only about 30 miles per day, and therefore should not recoil at the prospect of a limited-range car, and econometric studies that show those same motorists would have to be compensated over $10,000 to accept a limited-range vehicle.

We interviewed households with at least two cars and a convenient place on the premises for recharging. One of the vehicles had to be a compact or subcompact purchased in the past two years. Each driver was asked to complete a detailed trip diary for one week. The trip data were collected and organized for easy review and presentation to the entire household. Interviewers explored how a limited-range vehicle might meet some of the household's weekly travel needs. Household members reconstructed their weekly diaries under a variety of scenarios—for instance, parents having to rush a child to the hospital—modified only by the substitution of a hypothetical electric vehicle. They were asked how they might be willing to adapt, if at all, to range constraints, and what minimum range they would accept. Because of the expense of conducting these interviews—almost $2,000 per household—only fifty-one households in the San Jose, Sacramento, and Los Angeles areas of California were interviewed.

The results were surprising. We found that many households might be satisfied with less than a hundred miles of range (see Table 3-5). Half the households indicated that fifty miles or less was all they needed for one of their cars—in most cases the recently purchased subcompact or compact car—and that ninety miles would suit almost all circumstances.[36]

We also found the argument that consumers would be unwilling to buy expensive electric vehicles to replace cheap second cars flawed. Today the differentiation between first and second cars is often meaningless, an anachronistic throwback to the postwar pattern of the nuclear family where Dad drove to work and Mom used a second car to do the shopping. Indeed, under scenarios that encouraged households to maximize their use of the electric vehicle—for instance, gasoline costing $5 per gallon—almost all assigned many more miles to the hypothetical electric vehicle

TABLE 3-5
Minimum Driving Range Desired by
51 California Households, 1994

Driving Range (miles)	Number of Households That Demanded at Least This Range	Number of Households Comfortable with This Range
40	14	1
50	12	3
60	8	5
70	3	3
80	6	10
90	0	3
100	2	12
120	2	6
Unlimited	4	8

Source: K. Kurani, T. Turrentine, and D. Sperling, "Demand for Electric Vehicles in Hybrid Households," Transport Policy (forthcoming).

than to the gasoline car it replaced. On average, they traveled 17 percent more miles in the hypothetical electric vehicle than in the gasoline vehicle, and one-fourth traveled 25 percent more. These households were thinking in terms not of first and second cars but of short versus long range.

Because fifty-one households is too small a sample to be representative, we streamlined our survey technique to build on what we had learned for a larger population sample. We recruited 600 households from around California, including Fresno, giving each fifty dollars so that not all respondents would be electric vehicle enthusiasts. Again, only households with two or more vehicles were interviewed. First, we sent each household a questionnaire to obtain basic information on its demographics, environmental attitudes, and intentions regarding its next vehicle purchase. Then we asked respondents to complete a three-day trip diary for each vehicle they owned. Next we mailed a specially prepared video to give them some sense of what an electric vehicle is, how it performs, and how it is recharged. In the last mailing we asked respondents to complete another questionnaire in which they were given various choices of vehicle, including different sizes and driving ranges. They were told that the EV would cost about $4,000 more than a comparable gasoline car, but that about $4,000 in tax credits and rebates would be available.

Preliminary returns reflected an even more positive response to limited-range electric vehicles than before. Of the total, only 54 percent of households chose gasoline vehicles. Sixteen percent chose electric vehicles with ranges of 80 miles, and 30 percent chose electric vehicles with ranges of 100 miles.[37] (The 100-mile electric vehicle was posited to cost $800 more than the 80-mile one.) Of these multicar households, then, almost half were willing to buy electric vehicles with ranges of 80 to 100 miles.

We concluded from these two studies that limited range would be a relatively minor drawback for many households, and more than offset by the prospect of home recharging, "greenness," and lower operating costs.

This is not say that battery-powered electric vehicles would suit all households. Many people travel extensively and erratically, often because they use their vehicles for sales trips or other work-related activities. It is people who do not usually drive far from home in any given day, who have routine driving patterns, and who buy new cars regularly that comprise the initial target market.

Much more needs to be understood about driving and purchasing habits before we can predict the potential market for electric vehicles. Virtually nothing is known, for instance, about vehicle swapping among household members (to what extent cars are treated as strictly "his" and "hers"), how frequently households take long trips, how willing they are to rent vehicles, and what value they attach to home recharging.

Inventing an Electric Vehicle Industry

Conventional car companies are based on a highly capital-intensive industrial model in which an assembler firm strives to control supplier firms further down the pyramid. Sales and services within this system involve large networks of independent dealers with massive investments and inventories, making most of their living off repairs and replacement parts. A successful electric vehicle industry is unlikely to resemble or spring from such an industry, claims James Womack, a leading expert on the auto industry.[38]

Why is a new organizational format needed for an electric vehicle industry? First, electric vehicles will differ from conventional vehicles in performance and in producing less noise and vibration, having shorter range, and requiring less maintenance. A radically different type of marketing and distribution system would support the vehicle and explain it to new buyers.

Second, electric vehicles will have different body materials and motive power, and they will be produced initially at lower volumes than conventional cars. This suggests the need for new approaches to manufacturing. One estimate is that 30 percent of the parts of a conventional car will be modified in an electric vehicle, and 40 percent will be entirely new.

Third, the high cost of batteries and recharge systems, along with rapid innovations that will depress resale value, calls for new ways to produce, distribute, finance, and service electric vehicles. A new infrastructure would reduce capital investments and overall costs, as well as consumer risk.

According to Womack, there are two types of companies that might be successful in an electric vehicle industry: a "mobility provider" that owns and services electric vehicles for the consumer; and clusters of "lean" manufacturing companies, organized horizontally rather than vertically, as is the case with the assembler-supplier groups prevalent today.

The mobility provider would order a vehicle directly from a manufacturer (or manufacturing group) and rent it to the consumer at a fixed monthly cost plus a variable cost based on driving. The mobility provider would obtain insurance, register the vehicle, service it at the owner's home, and replace it on request with a new vehicle. This unique service would also be desirable with a conventional vehicle, but the high initial cost of electric vehicles makes the concept more urgent and low maintenance makes it attractive. By renting, consumers would be free to replace their electric vehicle if and when range and carrying capacity requirements change. Costs would diminish with the elimination of large showrooms, inventories, and repair facilities that conventional dealers maintain. The "mobility provider" might evolve from a car rental company or an unregulated subsidiary of a local electric utility.

The manufacturing side of the industry might take the form of long-term collaboration among company clusters rather than the hierarchical assembler-supplier arrangement. For example, an aluminum or structural plastics firm might provide the frame, an electronics firm the powertrain, and other plastics firms the exterior skin and interior fittings. A traditional car parts firm could engineer and produce the suspension. The mobility provider that took on marketing and customer support in a given geographic area might affiliate with one or more assemblers. Today's assemblers might be replaced by small companies of several hundred employees who oversee integration of the vehicle system and assemble large subsystems provided by other companies. This new structure might resemble trends in today's computer industry where small assemblers package components for generic computers, and "computing providers" rent computer services, not specific computers, to businesses.

The electric vehicle industry of the future might evolve differently. But if it is to succeed, it must, as Womack argues, be different from the current auto industry. The seeds of a new electric vehicle industry can be found in Calstart, a loose consortium of companies and governments in California. Created in 1992, it has spawned a network of businesses that collaborate in designing and producing compatible components for electric vehicles. It has also established links between government agencies and businesses, resulting in the funding of technical training programs for workers and students and a variety of grants and subsidies for those companies. Whether Calstart will mature beyond a consortium of component suppliers into a Womack-style horizontal electric vehicle industry remains to be seen.

Although the immediate future of electric vehicles is highly uncertain, conditions seem to point in their favor. First, as the most environmentally attractive transportation option available, they are likely to receive extraordinary government support. The major near-term environmental benefit they offer is improved air quality, the form of government regulation perhaps most enthusiastically embraced by the public and among the most far-reaching. Second, electric utilities support the electric vehicles because it will allow them to draw power from otherwise idle capacity. There is no need to build new capacity to support large numbers of electric vehicles; thus they can reduce overall system costs per unit of electricity sold. Third, electric vehicles don't require a new fuel distribution network, as do vehicles fueled by methanol and CNG. Electricity is available virtually everywhere, and the small cost of setting up a charging station in homes and businesses would be at least partly covered by utilities.

Because of the initial high cost of buying an electric vehicle, the first owners would probably be fleets and affluent consumers who have a strong commitment to the environment. With technology improving and vehicles being produced in ever-larger numbers, costs would steadily diminish. Consumer perceptions and behavior, frozen by a century of gasoline technology, would thaw as market penetration expanded. Just as they were surprised at the strength of consumer demand for air bags, automakers might well discover that they underestimated demand for electric vehicles and the advantages they have to offer.

There is no substitute for government subsidies and incentives to jump start the electric vehicle industry in its initial stages, when vehicles look to be costly and unfamiliar to consumers. Given the huge energy and environmental benefits of electric propulsion, government intervention is justified; indeed, one could argue that the government is morally obligated to intervene. The challenge for government and industry is to create a positive culture for electric vehicles—by reducing uncertainty and risk for the vehicle buyer, and by demonstrating a strong, stable commitment to electric vehicles as well other environmentally attractive alternative-fuel vehicles.

CHAPTER 4

Neighborhood Electric Vehicles

A Neighborhood of the Future

Over the past 5 years not a single person has been seriously injured by a car in this neighborhood, even though it is heavily traveled by pedestrians and bicyclists. The speed limit is 20 mph for all vehicles, including trucks and large cars, whose speed controls are activated when they leave major roads and enter local residential and commercial streets. Most people live in single-family homes with yards, but garages and driveways are much smaller than in the 1990s and the streets are only half as wide.

What's most striking in this suburb of the future, though, is not the safety record, the speed limit, nor the size of driveways and streets. It is the large number of small, colorful cars known as neighborhood electric vehicles, NEVs for short. Residents of this community still drive gasoline cars on occasion, but they pretty much stick with their NEVs. They take as many trips by car as people did in the 1990s, but the trips are shorter and the people walk and bicycle more, a trend that has led to the revival of neighborhood shops.

Let's drop in on a typical family of five that owns four vehicles—three small electric "runabouts" and one gasoline-powered minivan. The van is parked at a public garage a few blocks away and driven only on long weekend trips. The electric cars are the everyday vehicles. One is used by the homemaker to run errands and drive a preschooler to day care. One belongs to the family's fourteen-year-old, who drives it to after-school activities. The family breadwinner drives a narrow electric commuter car to the office eight miles away.

The couple next door own just two electric cars. Rather than own they prefer to rent a gasoline vehicle for their occasional weekend trips. The zippy sportscar they like is waiting for them at the public garage when they request it. The cost is minimal, since they received five free vouchers a year with the purchase of each of their electric cars. The vouchers cover over half the rental costs for the year.

The neighbors on the other side, who are elderly, paid $1,000 extra for an electric car with automated controls for steering, accelerating, and braking, and $300 extra so it can be operated from a wheelchair instead of the standard driver's seat. The wheelchair rolls into the driver or passenger side of the car and locks into place.

Moderate-sized trucks delivering large boxes are allowed to enter this neighborhood from ten in the morning until noon. A trucking company pays a $30 fee for the privilege of using the road. Or truckers can deliver and pick up goods at a privately owned terminal, licensed by the city, at the edge of the neighborhood. There the goods are transferred to small electric trucks that serve residences on the narrow street. Businesses in town make deliveries and pickups in their own small electric trucks. Emergency vehicles such as firetrucks and ambulances, downsized, are given priority access to the neighborhood street.

Brave new world? Utopia? It is neither, but it may be a workable solution to problems afflicting many localities: too many cars taking up too much room and traveling too fast. Today's powerful cars can carry four or more people in comfort, accelerate quickly to 60 mph, and cruise comfortably at 70 mph. This broad capability is costly in terms of fuel and space, not to mention indirect costs of environmental damage and maintaining an expensive, wasteful transportation system. And the fact is that large, full-powered vehicles are not even necessary for most trips. About half of all car trips cover less than 5 miles and are taken by a single person traveling at generally low speed.[1] Continued attachment to large cars is impeding efforts to reduce energy consumption, adopt battery-powered, zero-emission vehicles, and design more pleasant neighborhoods.

The problem with our transportation today is threefold: all vehicles are expected to satisfy all purposes; all roads are built to serve all vehicles; and all rules are designed for the standard vehicle of the past. The key to introducing small cars is dispensing with this one-size-fits-all mentality. Changes are needed in rigid safety regulations that stifle innovation, in traditional manufacturing methods that discourage small cars, standardized infrastructures that discriminate against small vehicles, and traffic control rules that serve only large vehicles.

A principal force for change is increasing affluence and multiple-car ownership. Almost 40 percent of American households own two vehicles and an additional 20 percent own three or more, and the percentages are increasing.[2] This means that no longer must every car serve every purpose—cars can be designed to suit specialized needs. It is happening already with the upswing in sales of small and large vans, two-seater sports cars, and luxury four-wheel-drive sport utility vehicles (such as the Jeep Cherokee).

The stage is set for a family of new vehicles smaller than subcompact cars: narrow cars for commuting at high speed on special freeway lanes; shared "station" cars to travel to and from transit stops and stations; small trucks for college campuses, office parks, military bases, and crowded downtowns; and neighborhood cars used strictly for local travel. These mini "niche" vehicles, taking up less road and parking space and consuming less energy, could eventually constitute a substantial share of the total car market.

Neighborhood electric vehicles, or NEVs, offer some of the most exciting possibilities. NEVs would be small, light vehicles built specifically for neighborhood trips, not for traveling on freeways. Most would have room for one or two people plus storage space, while some would be larger to accommodate families with several children. The cars, traveling no faster than 40 mph, would typically be capable of making trips of 30 miles or so before recharging.

Station cars are a variation of the NEV concept. Shared by those driving to and from transit stops, these small, jointly owned station cars would also be cheap. The vehicle might be dropped off at a station by one individual, picked up by another person arriving at the station, and perhaps even returned and used again in the evening by the first driver. Station cars might also be used for personal errands and business trips during the day.

Europe and Japan have had limited success with experiments in station cars. In the United States, several transit operators and electric utilities recently formed a station car association. They began purchasing vehicles and making them available to volunteers in 1994. Initial vehicles are subcompact-sized, but smaller neighborhood ones will appear as the program expands and the vehicles become readily available.

NEVs are not golf carts. Low-end versions, such as a prototype being readied for production by the small Michigan company Trans 2, are similar in size and carrying capacity (two seats) to golf carts, and only a little faster (22 mph versus 15 mph), but they offer superior performance, safety, and comfort. The Trans 2 has a lower center of gravity, front-wheel drive, and carlike suspension for improved stability, cornering, and maneuverability. Unlike a golf cart, it comes with windshield wipers, a horn, side-view mirrors, and three-point seat belts anchored to the frame. The vehicle has a higher and more visible profile than a golf cart, a full array of gauges, and lockable storage areas. The Trans 2 has a range of 25 miles and a target retail price of $5,000.

Several larger NEV types are manufactured in Scandinavia. The three-wheel City-El accommodates one person plus storage space and travels up to 30 miles on one charge at a speed of up to 35 mph. It is made in Denmark, but negotiations are proceeding for a subsidiary in California. About 5,000 were sold in the 1980s and early 1990s in Europe; the price is $7,000, including batteries. The Kewet, also produced in Denmark, was first sold in the United States in 1993. It has four wheels, a top speed of 38 mph, a range of 30 miles, two seats, space for small packages, and costs $12,800. Both the City-El and the Kewet are essentially hand built with primitive technology. If the vehicles were mass-produced in a modern factory, the cost would probably be cut in half.

Neighborhood vehicles need not be electric.[3] They could burn gasoline or some other fossil fuel in an ICE and still be as cheap or cheaper to the consumer than a pure battery-powered neighborhood vehicle. However, since the principal reason for developing NEVs is the automakers' need to meet ZEV mandates, battery power will probably predominate.

The Trans 2 prototype from Michigan.

At a neighborhood electric vehicle exposition at the University of California, Davis, are (*left to right*) the City-El from Denmark and a Horlacher, an Esoro, and another Horlacher, all from Switzerland.

The Kewet vehicle from Denmark.

Moreover, as NEV technology improves, mini–electric cars will undoubtedly be seen as superior to mini–gasoline cars in convenience and reliability, if not cost.

The Energy and Environmental Benefits

NEVs are attractive because they consume much less energy and emit (at the power plant) much lower levels of greenhouse gases and other air pollutants than larger gasoline and battery-powered vehicles. (This would be true even if power plants supplying the electricity relied primarily on coal.) NEVs would be used for short, slow urban trips—where gasoline engines are at their most polluting and inefficient. Emissions from gasoline engines are ten or more times higher during short trips and in the first few minutes of longer trips when catalytic converters are still cold. By contrast, NEVs and other electric vehicles have no tailpipe emissions or catalyst warm-up problem. As a result, NEVs have the same low emissions from the power plant in the last mile as in the first.

The huge energy savings and environmental gains to be had with NEVs are clear if we compare the performance of a Kewet with that of a subcompact gasoline car. Assuming trips averaging 2.5 miles, speeds varying between 0 and 35 mph, about 60 percent of trips being cold start, and electricity for the Kewet coming from an average mix of U.S. power plants (52 percent coal), the Kewet would emit 99 percent less carbon monoxide than the gasoline car, 99 percent less hydrocarbon, and 92 percent less nitrogen oxides.[4] The Kewet would also consume 50 to 60 percent less energy and emit 50 to 60 percent less greenhouse gases than the gasoline-powered subcompact.

Land Use and Mobility

NEVs could also enhance mobility for many people. It is estimated that over 10 million persons of driving-age in the United States have a physical disability that makes them dependent on sparse public transit services or expensive specialized services. This number is likely to increase as the baby boom generation enters old age. Between 1990 and 2020, the over-fifty population is expected to almost double, from 63 million to 112 million. The ease of driving a NEV makes it accessible to a broader range of individuals, including those with physical disabilities. NEV driving could be made even easier by incorporating fully or partially automated controls, further expanding the number of people with access to personal transportation.[5] Automated controls are used today in slow vehicles in controlled settings such as factories, but an intense effort is under way to incorporate controls into other vehicles. Such controls are much more easily accommodated to NEVs than to vehicles operating at faster freeway speeds.

Another important effect of NEVs might be, paradoxically, to encourage a return to walking and bicycle riding. These activities are not for everyone, but just a small shift away from motorized vehicles would save energy and reduce emissions. Even in small numbers, NEVs could force local governments to rethink the auto-centric planning that predominates in many locales today by redirecting funding, building and traffic rules, and road and land development investments. As city planners come to encourage bikeways and pedestrian paths, these roads and paths would become more widespread and intensively used.

In fact, this sort of change has already taken place without the prod of NEVs. Consider the case of Davis, California. Through the 1960s, the city and the University of California campus had a typical gridlike, auto-based road system. A few individuals started a campaign to promote bicycles. Through trial and error, initially against the opposition of local officials but later with their support, roads and traffic controls were gradually adapted to bicycle use. Some roads were closed to motorized vehicles, others required special permits, bicycle lanes were created on still others, and eventually entirely new bicycle paths were built. The city and campus are now covered by an integrated network of bicycle paths and lanes, with traffic circles and other traffic-control devices designed specifically for bicycles. Police officers on bikes enforce traffic rules, including observance of stop signs, and issue tickets to offenders.

The same thing could happen with NEVs. While guidelines can and should be developed to assist local planners and officials, each community would need to grapple with its own peculiar set of circumstances and obstacles. As in Davis, some roads might be closed to conventional vehicles permanently or during certain hours, with narrower (and cheaper) roads such as cul-de-sacs built for NEVs. Many inexpensive infrastructure changes could be made to accommodate and even reward NEV use.[6] On wide roads, governments could set aside lanes for NEVs and create special crossing areas at major arteries. In communities with transit stations and park-and-ride lots, there could be special parking for NEVs; in shopping areas and workplaces, preferential parking.

Changes in road design are as needed as they are difficult to bring about. Today's municipal engineers and planners build wide streets that are empty most of the time. Guidelines call for a minimum street width of 22 feet, but roads are usually built much wider, and developers who want to build narrow roads have to go through an arduous appeals process. Standardized safety provisions and roadway designs were created with conventional-sized vehicles in mind. One might argue that the road system should be designed to serve pedestrians, bicycles, NEVs, conventional cars, and service trucks, in that order. It would look very different from the road systems in most suburban communities.

The pressure to reduce driving and resurrect neighborhood communities is reflected in the popularization of neotraditional urban planning, with its emphasis on clustered development, pedestrian travel, and mass transit.[7] The reality, though, is that many people are unable to walk or bicycle, or don't want to in cold or rainy weather, or balk at the inconvenience and invasion of privacy associated with the shared vehicle, or live in areas poorly served by fixed-route, fixed-schedule conventional transit. Whether clustered and pedestrian-friendly development makes a strong comeback may well depend on the NEV.

NEVs would provide travelers with choice, a crucial political condition for new regulations. The experience of the late 1980s with alternative fuels and emission standards is instructive. Air quality regulators had been reluctant for many years to impose sharp new emission reductions on cars, trucks, and buses. Not until clean-burning methanol and natural gas were perceived as plausible alternatives to gasoline and diesel fuel were regulators emboldened to impose much more stringent standards. In the case of national truck and bus standards and California car standards, regulators maintained that, if necessary, manufacturers could always fall back on alternative fuels to meet requirements. NEVs could play the same catalytic role in land-use management and infrastructure redesign that clean fuels played with vehicle emission regulations in the late 1980s.

The ZEV Mandate as an Instrument of Change

The ZEV mandate could be the main instrument for introducing neighborhood electric vehicles. As automakers confront the high cost of meeting the ZEV mandate with full-size electric vehicles, they may well become more receptive to the concept of the NEV, which is smaller, less costly, and easier to build. NEVs are arguably a more compelling application of battery-powered electric propulsion than full-size vehicles because

NEVs—light, low speed, easier to cool and heat, and designed for short ranges—need much less power and use much less energy.[8]

The low energy needs allow NEVs to run on small battery packs. The City-El, for instance, has a 240-pound conventional lead-acid battery pack, the Kewet a 590-pound pack, the Trans 2 a 280-pound pack. These vehicles represent relatively unsophisticated engineering. Future NEVs should be more energy efficient and carry even smaller battery packs. A typical subcompact electric vehicle, by contrast, carries a (lead-acid) battery-pack of over 1,000 pounds. Even the very energy-efficient Impact prototype carries 900 pounds.

Because batteries account for a large share of the total cost of electric vehicles, the reduced need for batteries improves the price tag of NEVs. Under conditions of mass production, the NEV should prove much cheaper to own and operate than a full-size gasoline or electric car.

The seven major automakers affected by the ZEV mandate could take advantage of the relatively low cost of NEVs in one of two ways.[9] They could make and sell NEVs, an attractive option because per-vehicle losses would be much lower than those associated with conventional electric vehicles. Initially, the loss per conventional electric vehicle could be $5,000 or more, depending on the availability of government incentives and subsidies. By contrast, NEVs should not cost more than $10,000 total. The second option for automakers would be to forgo the manufacture of ZEVs and to buy ZEV credits from NEV manufacturers, most of whom will probably be too small to be subject to the ZEV quotas (with the possible exception of some Japanese companies). Purchasing ZEV credits for NEVs would be much cheaper for the Big Seven than building and selling full-size electric vehicles, and it would provide additional revenue to NEV manufacturers. Given that the fine for not producing a required ZEV is $5,000 per vehicle, the value of a ZEV credit for each NEV (or any other EV) could be anywhere up to $5,000 (for a full description of the ZEV mandate, see pp. 138–140).

The California Air Resources Board has blessed NEVs implicitly by determining that all four-wheel vehicles registered for road use qualify as ZEVs for purposes of the mandate. Currently, three-wheel vehicles don't count in the ZEV mandate's total tally because they are registered as motorcycles. In the future, if and when NEVs become common, regulators may want to re-open the issue of NEVs receiving full ZEV credit. Should, for instance, four-wheel vehicles not capable of freeway speeds and operating on separate roadways count as ZEVs in meeting the mandate requirements? A sensible solution may be to provide partial credit for vehicles according to their expected amount of use.

Safety and Liability

The safety of NEVs is possibly the most critical issue in determining how, where, and when NEVs will be successfully introduced to American consumers. Safety regulators in the United States are diligent, determined, and effective. Their mission, to make individual vehicles safer in case of accident, frames safety debates. But this approach to safety is narrow. It misses the larger benefits that result from viewing safety as a systems problem. It ignores pedestrians, for instance, who account for about 15 percent of all traffic deaths in the United States. Wouldn't safety be enhanced if speed were reduced in neighborhoods and local commercial districts, and access by large vehicles restricted? Unfortunately, there is little evidence available on which to base a realistic assessment of the overall safety hazards and benefits associated with a NEV-based transportation system.

The debate over NEV safety will probably be narrower than it should be and may well focus on the undeniable physical reality that an occupant of a small car is clearly more vulnerable to injury than an occupant of a larger car, all else being equal. But all else need not be equal. Trucks could be banned on roads designated for NEVs, and the speed limit could be reduced on NEV roads, using speed bumps and other traffic-calming devices.

Small cars could be made safer through better design and with safety devices. For instance, a Swiss company called Horlacher has designed several very small prototype vehicles that are stiff and use air bags and seat belts. In crash tests, dummy occupants of a Horlacher vehicle reportedly received fewer injuries in a collision with an Audi than the Audi occupants. An ultra-stiff shell with internal restraints is what allows race car drivers to survive crashes at speeds in excess of 150 mph.

There is an important unanswered question: What type of safety standards should be developed for neighborhood electric vehicles? Through the National Highway Traffic Safety Administration (NHTSA), the federal government has established the most detailed and prescriptive vehicle safety standards in the world. However, none of these is appropriate to the vehicles we have been discussing. Currently there are no safety regulations or laws specific to electric vehicles of any type. Several proposals are under consideration as this book goes to press, but these deal with recharging, crash avoidance, and crash worthiness. There appears to be little interest in promulgating standards specific to small vehicles. For now, NEVs must fit into the existing mosaic of rules and regulations for light-duty passenger cars and trucks, motorcycles, and golf carts.

In some states, NEVs might be put in the same category as less-regulated motorcycles. The federal government defines a motorcycle simply as a vehicle with two or three wheels, but each state has its own definition. California uses the more stringent definition of a motorcycle, that is, a vehicle weighing less than 1,500 pounds, unless it is electrically powered, in which case it can weigh 2,500 pounds but not exceed 45 mph.

Golf carts are not classified by the federal government. California defines them as vehicles weighing less than 1,300 pounds, operating at 15 mph or less, and carrying golf equipment and not more than two persons. Arizona, by contrast, allows golf carts to be registered as recreational vehicles and licensed as motorcycles. The only requirement there, in addition to licensing and registration, is that they not impede the flow of traffic.

As an indication of the confusion, note that the three-wheel City-El is registered in California as a motorcycle. Because motorcycle safety standards are far less stringent than car standards, the City-El and other prospective three-wheel NEVs in California and elsewhere currently face no significant regulatory barriers.

One disadvantage of a motorcycle designation is that most states require motorcycle users to wear a helmet, a disincentive to buyers. The helmet law could be waived for three-wheel NEVs, but in most states, including California, that would require legislation.

Safety standards present more of a problem for four-wheel NEVs, even though they are safer. They must meet the same standards as a full-size vehicle, even though they do not travel on freeways or at high speeds. There have been exceptions. In 1967 a broad exemption from the standards was granted for four-wheel vehicles under 1,000 pounds on the grounds that it was impossible for such vehicles to meet the general standards. The exemption was removed in 1973, and many subsequent failures to reinstate a similar exemption reflect NHTSA's insistence that all vehicles meet the same standards.[10] How difficult it would be to obtain a permanent exemption or to create a new category for NEVs is uncertain.

Manufacturers are left with two options. They can petition for an amendment to any impractical standard, or they can apply for a temporary exemption. Amendments are difficult to win. Exemptions can be granted for up to 2,500 vehicles per year to manufacturers who build fewer than 10,000 vehicles a year on any of the following grounds: substantial economic hardship, aiding the development of new vehicle safety or low-emission engines, or providing an equivalent level of safety. A NEV would easily qualify on the low-emission criterion, and possibly on the other two as well. The exemptions are renewable, but it is not clear how many renewals would be granted.

Liability could prove to be a larger concern for NEV manufacturers.

The promulgation of safety regulations would provide some legal protection to NEV manufacturers, but the small size of NEVs will still raise safety questions.

There are three types of product liability: negligence, warranty, and what is called strict liability. The first two are not a concern, as there are no new or unique negligence or warranty issues associated with NEVs. But strict liability is another matter. It covers manufacturing, design, and warning defects. The question of defects is fuzzy. For instance, three-wheel vehicles with the single wheel in front are often considered more dangerous than those with a single-wheel-in-back configuration. Would an automaker offering the first type be liable under the terms of strict liability?

In determining liability risk for NEVs, the pivotal question is this: Does the use of a NEV pose an unreasonable danger to the user? Precedent holds that a danger is unreasonable only if it is not clear and obvious to the user. Manufacturers of bicycles and motorcycles are protected from claims because these vehicles are clearly dangerous. Being aware of the danger, a driver implicitly accepts the risk. It is likely that makers of NEVs, especially low-end NEVs, would be protected under the same umbrella.

NEV manufacturers would find additional protection if their designs were determined to be state-of-the-art manufacturing at the time of production, but this sort of liability determination—as well as most others—is highly subjective. A goal, from the perspective of both safety and liability, might be to create designated areas for NEVs, such as drive-slow zones. A NEV involved in an accident in such a zone would not be at fault, just as pedestrians on crosswalks are not.

Whether NEV manufacturers would be burdened with liability claims remains unknown. One product liability expert thinks not; speaking at a workshop on sub-cars, he concluded that NEVs pose no greater liability than any other vehicle so long as appropriate effort is made to minimize risk.[11] In any case, liability concerns would lessen if federal safety regulators were to address NEVs. The sooner the better. Until then, industry will shy away from the NEV market.

Palm Desert Pioneers

Leading the way in incorporating small electric vehicles into the local transportation system is Palm Desert, California.[12] In this case, the vehicles are simple golf carts—considerably less safe than NEVs.

Until recently in California, golf carts were permitted only on streets with speed limits of 25 mph or less and within 1.5 miles of a golf course. At the request of the city of Palm Desert and other communities, California's attorney general wrote an opinion in 1992 stating that golf carts could be driven on *any* street with a speed limit of 25 mph or lower, but not on or across any street with a higher posted limit. Based on a survey of residents indicating that many people in this small, affluent resort community would use their golf carts for local travel, if allowed to do so, Palm Desert sought to create a golf cart demonstration program in 1993. Because of opposition from the California Highway Patrol and a desire to use higher-speed roads, an act of the state legislature was required to authorize the demonstration program and define the limits under which electric golf carts would be allowed to travel public streets. With support from the South Coast Air Quality Management District and the local state representative, the enabling legislation was passed. An oversight committee, consisting of state officials along with representatives of the city, air quality district, California Highway Patrol, and local sheriff's office, was set up to monitor progress.

The program, intended to demonstrate the potential of small, low-speed vehicles to improve air quality, rules out gasoline-powered carts. Each cart must be equipped with basic safety equipment such as a horn, turn signals, and headlights and must pass a city inspection, while the owner is required to take part in an orientation meeting. Upon completion of this process, the cart is issued a permit by the city allowing it to travel on designated right-of ways within Palm Desert.

A golf cart being used for transportation on a specially marked lane in Palm Desert, California.

A three-tier system of right-of-ways was developed for licensed golf carts. They are allowed to travel in mixed traffic on any street with a speed limit of 25 mph or less, in accordance with the attorney general's opinion; in separate designated lanes on certain streets with higher speed limits (in some places, these lanes are shared with bicycles); and on golf cart paths completely separated from vehicle right-of-ways (several such paths have been built, with plans in the works for additional ones). Meanwhile, special traffic signs and signals are being refined to educate golf cart and motor vehicle drivers about the new infrastructure.

Palm Desert is an object lesson in how NEVs could be accommodated in local communities. What is learned there and in other locales experimenting with small vehicles will be valuable when it comes time to design and regulate NEVs for the communities in which they will operate.

Creating an Industry

The major car companies have not been eager to build neighborhood vehicles, not only for fear of safety related liability but also because automakers, especially in the United States, enjoy much greater profits from large vehicles produced in large volume.

The fundamental problem is that car manufacturers are still wedded to the mass-manufacturing methods developed by Henry Ford and based on the use of steel. Today the economics of steel production forces rigid, high-volume manufacturing that involves a phalanx of dies costing upward of $1 billion per vehicle model to shape the steel; an electrocoating plant that costs $0.25 billion; and a $0.5 billion paint shop.[13] Only about 15 percent of a steel part's cost covers the steel; the other 85 percent has to do with shaping and finishing the raw material.[14]

The steel-inspired approach to car manufacturing has been remarkably successful in reducing cost, but it cannot exploit the opportunities presented by lightweight materials, and it isn't conducive to small production runs. Automakers have adopted lightweight nonmetallic materials in recent years, but as straightforward substitutions for steel parts, with no change in the manufacturing process.

The situation is far from hopeless. The trend already under way toward specialized vehicles, the advent of so-called flexible manufacturing, and the willingness of smaller companies to enter the industry all enhance prospects for NEV manufacturing. Flexible manufacturing uses computers to control manufacturing. Instead of repetitively stamping and assembling hundreds of thousands of nearly identical parts and vehicles, programmed computers direct machines to substitute coded items on certain batches of vehicles. This reduces the cost per vehicle for smaller production volumes and makes it feasible to produce more vehicle types in a single factory.

Even more intriguing is the possibility of manufacturing vehicles in entirely new ways using composite materials and lightweight construction. Amory Lovins, a prominent long-time advocate of energy efficiency, argues that by shifting from steel to composite and other lightweight plastic materials, and revamping how vehicles are manufactured, much lighter vehicles can be produced in smaller lots that are just as safe as conventional vehicles and no more expensive.[15] If he is right, it would be a boon to small electric vehicles such as NEVs.

The problem today is that small cars cost almost as much to manufacture as large cars, and must be produced in great quantity to be competitive. Under Lovins's proposal, production would no longer be beholden to the high fixed cost of steel-based manufacturing, and thus the need for large economies of scale would diminish. Smaller vehicles produced in smaller quantities would cost much less than larger cars.

This revolutionary new approach to manufacturing is premised on the strength of carbon fibers and the ease with which composite materials can be shaped. The materials allow a dramatic reduction in the number of parts—2 to 6 for a car body (excluding powertrain, trim, and chassis) versus 300 to 400 using traditional materials and methods—and therefore a sharp decrease in assembly time and cost. Composite materials are also highly durable and rustproof, they don't need the expensive treatments given to metals, and they can be painted more easily and safely. Though they are much more expensive than metal, the cost could be largely offset by other savings.

This new approach to manufacturing has not yet been fully tested, but with the cost of composite materials continuing to drop, it could make gains. It is especially promising for NEVs—because of the importance to those vehicles of weight reduction, because smaller companies that lack the capital for mass-manufacturing plants could enter the industry, and because only a small quantity of each model would likely be sold at the outset.

Major American automakers are decidedly more reluctant to enter the NEV market than automakers elsewhere. Most Japanese vehicle manufacturers, especially the smaller companies, sell very small gasoline cars at home and elsewhere, and some European companies sell very small cars in Europe. In 1993, 1.6 million minivehicles—vehicles with engine displacement no larger that 660 cubic centimeters (cc)—were sold in Japan by Suzuki (which also makes GM's Geo Metro), Diahatsu, Honda, Mazda, Mitsubishi, and Fuji. New companies, or companies more flexible than is typical in the tradition-bound auto industry, will probably be most successful in exploiting NEV production and marketing opportunities. A variety of nonautomotive and smaller companies are interested in the NEV market—manufacturers of composite and fiberglass materials such as Horlacher in Switzerland, start-up companies such as Trans 2, and builders and converters of specialized vehicles, such as those that put convertible tops on cars. Interestingly enough, Swatch, the Swiss watchmaker, is collaborating with Mercedes-Benz to produce a mini–gasoline vehicle in the late 1990s and an electric version shortly thereafter.

In practice, small and nonautomotive companies would probably seek alliances with larger companies out of the need for capital, marketing expertise, and a distribution infrastructure. It is beyond the financial and organizational capability of small companies to create and operate a network of exclusive dealerships. Alliances could range from simple marketing assistance to outright purchase and resale of vehicles by a giant automaker, already a common practice.

The Market for NEVs

What are the long-range prospects for sales of NEVs? The market in the United States alone might be millions per year. Even in the short term, with little change in consumer expectation or government regulation, demand might reach a sizable level. But credible quantitative estimates of the potential market cannot be made at this time. Research is fragmentary and speculative. According to confidential industry marketing studies, about 140,000 golf carts and small electric industrial vehicles are sold annually in the United States. One study estimates that about 20,000 golf carts are used at least in part for personal transportation. Market penetration will depend on a large number of factors related to ZEV and safety rules, local initiatives to accommodate NEVs, liability law, traffic control regulations, and the entrepreneurial initiative of manufacturers.

We've already noted some markets for NEVs: mobility-impaired individuals (estimated at about 10 million people in the United States), factories, and commuters driving short distances to urban rail and bus stations and park-and-ride lots. The market is also likely to include resort facilities, national parks, and owners of the approximately 3 million vacation homes in the United States, where clean and uncongested landscapes are highly valued. Many resort communities have already imposed restrictions on the use of cars, and the trend is for further restriction. In heavily visited park areas, such as Yosemite, that are showing signs of damage from exhaust, a sensible strategy would be to ban gasoline and diesel vehicles and replace them with electric buses, electric cars, and NEVs.

Closed neighborhoods and communities with more restrictive speed limits are likely to be receptive to NEVs. Palm Desert, California, in addition to designating certain streets and lanes for electric golf carts, has also provided preferential parking in a number of locations. Many other communities are hospitable to golf carts, but less formally. Golf carts are widely used for transportation in a number of retirement communities in southern and western states.

Yet another potential market consists of large new developments that can be designed specifically for NEVs. In California alone, neighborhood electric cars are being considered as integral elements in four new town developments together covering 100,000 acres. Several developers are considering the idea of providing a NEV with some or all homes sold.

Neighborhood cars promise huge social, economic, and environmental benefits. They are energy efficient, low-polluting, and neighborhood friendly. The only possible drawback, the safety of occupants, can be mitigated, and it is conceivable that overall safety including that of pedestrians would improve with an NEV-based transportation system. The widespread use of NEVs could prompt lower speed limits in neighborhoods, while increasing mobility for the elderly and those with minor disabilities who otherwise would be unable to drive. NEVs would have the added advantage of being economical, in part because they are an ideal application of battery-powered electric propulsion.

The market for NEVs is potentially large but highly uncertain. Studies are needed to determine who would buy these vehicles, under what conditions, and with what incentives. With improved design, increased production, reduced cost, and reorganized institutions and regulations, the market share for NEVs should expand. They are clearly an attractive option that ought to be encouraged.

CHAPTER 5

Fuel Cells

We are at the very beginning of a new era of technology, comparable to the days when Gottlieb Daimler and Karl Benz were constructing the first vehicles powered by internal combustion engines.
   Helmut Weule, head of Daimler-Benz research and technology at their fuel cell vehicle unveiling in April 1994

Fuel cell vehicles seem almost too good to be true. They wouldn't require any fundamental changes in operation from conventional gasoline-powered vehicles. They could be refueled in minutes, their driving range would approach that of gasoline cars, and acceleration would be comparable. Fuel cell vehicles would consume about half the energy of gasoline-powered ICE vehicles, emit no pollution or greenhouse gases, and as icing on the cake, be quieter running.

Early models of the fuel cell vehicle could run on methanol or hydrogen made from natural gas, both rather easily available, or perhaps even a petroleum fuel. Eventually, fuel cell vehicles would likely be powered by hydrogen made from water using solar energy. At that point, the entire chain of activity from fuel production to driving would approach zero emission, and the dream of an environmentally benign transportation system would become reality.

What Is a Fuel Cell?

A fuel cell is a device that transforms hydrogen and oxygen into electricity. It is similar to a battery in that it has electrodes, an electrolyte, and positive and negative terminals. It does not, however, store energy as a battery does. Instead, fuel is supplied continuously, with the cell producing electricity all the while. This difference is a major advantage for the fuel cell system: it eliminates the problem of limited driving range. The separate fuel-storage container can be refilled from an outside source simply and quickly, in the same way that a gasoline tank is filled at a service station.

The vehicle's powertrain consists of a storage tank that holds hydrogen or a hydrogen-carrying fuel, a fuel cell system that converts the fuel into electricity, and an electric motor. A fuel cell vehicle may also include a battery, ultracapacitor, or flywheel for occasional power boosts (see pp. 105–107). The system that generates the on-board electricity includes one or more fuel cell stacks, plus auxiliary subsystems that, depending on the type of fuel cell, compress and supply air, cool the stacks, keep the insides of the fuel cell wet, and dispose of excess water. If the fuel is something besides hydrogen, a device known as a reformer will be needed as well for most fuel cell types to convert that fuel into hydrogen.

The electrochemical reaction in a fuel cell is far cleaner and more efficient than the chaotic combustion that takes place in an ICE. In an ICE, hydrogen and oxygen are mixed; a localized blast of energy (a "spark") causes oxygen and hydrogen to collide and release energy, most of which is wasted when the excited molecules bounce against the sides and top of the engine rather than against the moving piston. The excited hydrogen and oxygen molecules not only heat the engine and what surrounds it, they also react with other molecules such as nitrogen, forming nitrogen oxides and other undesirable compounds.

Fuel cell systems are at least twice as energy efficient as gasoline engines, while producing little or no waste heat, pollution, or noise. They keep hydrogen and oxygen separate. In fuel cells to be used in vehicles, hydrogen gas is piped to the negative electrode, or anode, as shown in Figure 5-1, and air is delivered to the positive electrode, or cathode. The hydrogen molecules separate, with the stream of electrons, or electricity, directed along a wire to the electric motor. The remaining hydrogen protons (positive ions) are pulled through an electrolyte and combine with oxygen that has been delivered as air to the cathode. The only products are waste water and useful electricity. The reactions are simple and easily controlled. Little energy is lost through heat.

The Fuel Cell Comes of Age

Although William Grove built a rudimentary fuel cell over 150 years ago in England, it wasn't until 1959 that a fuel cell vehicle was actually built. In that year Dr. Harry Karl Ihring of Allis-Chalmers Manufacturing, a U.S. company, constructed a fuel cell tractor. Because they were energy efficient, small, and nonpolluting, fuel cells were soon considered ideal for use on spaceships. The National Aeronautics and Space Administration (NASA) funded extensive research into fuel cells, which were used in the Gemini spacecraft in the mid-1960s and later in the Apollo spacecraft and space shuttles.

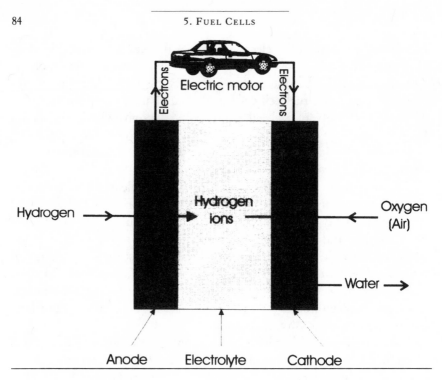

Figure 5-1 HOW A PEM FUEL CELL WORKS.

Out of the NASA experience grew a realization that fuel cells could be used to power automobiles. The initial burst of enthusiasm soon turned into disillusionment, however. Fuel cell technology, still primitive, did not fulfill expectation.[1] Fuel cells remained too heavy, bulky, and costly to be serious contenders for use in motor vehicles. The technology languished until adoption of the ZEV mandate in 1990. As research into batteries intensified and revealed their shortcomings, interest in fuel cells accelerated. Indeed, fuel cells were the impetus for the ambitious 1993 government-industry "Clean Car Initiative" (officially, the Partnership for a New Generation Vehicle) to develop cleaner and more efficient vehicles (see pp. 141–143).

By 1994, fuel cell vehicle demonstration projects were well under way in the United States and Europe.[2] The most advanced was a bus project developed by Ballard Power Systems of Canada in collaboration with several other Canadian and U.S. companies. The first bus, completed in June 1993, is a 32-foot transit vehicle powered by compressed hydrogen and a Ballard PEM (proton-exchange membrane) fuel cell. A second-generation bus using more advanced PEM fuel cells and compressed hydrogen, possibly to be supplemented with a battery for peak power, is scheduled for completion in 1996. Close behind are three identical fuel

The first fuel cell vehicle: a tractor built in 1959 by Allis-Chalmers.

A 5-kilowatt fuel cell stack made by Ballard Power Systems.

Ballard, SAIC, and others completed this fuel cell bus in 1993. The inset shows the PEM fuel cells and their auxiliary support equipment.

cell buses built under the supervision of Georgetown University with funding from the U.S. Department of Energy. The first bus was completed in April 1994. All three are powered by methanol, a phosphoric-acid fuel cell made by Fuji Electric, and a nickel-cadmium battery available for peak-power needs (for hard accelerations and climbing steep hills).

Similar projects are under way in Europe. The Eureka fuel cell bus, being developed by four European companies, has an alkaline fuel cell hybridized with a nickel-cadmium battery, and stores hydrogen on board in a cryogenic container. It is slated for completion in late 1994. Another bus, sponsored by the Commission of the European Communities, uses a PEM fuel cell hybridized with a lead-acid battery, and also stores liquid hydrogen. It is scheduled for operational testing in 1994.

Several smaller fuel cell vehicles are being built as well. In late 1993 Energy Partners, a start-up company in Florida, rolled out a PEM car running on compressed hydrogen. Daimler-Benz unveiled a van running on a PEM fuel cell and hydrogen in April 1994, and General Motors is building a methanol-fueled PEM car to be completed by 1996.

In mid-1994 the U.S. DOE announced contract awards of $13.8 million to Ford and $15 million to Chrysler to develop PEM fuel cells for light-duty vehicles. Both intend to use hydrogen as the on-board fuel.

This fuel cell bus, unveiled in April 1994, was built under the direction of Georgetown University with funding from the U.S. Department of Energy. It uses phosphoric-acid fuel cells.

Fuel Cell Technology

The four most advanced fuel cells for vehicles, classified according to the type of electrolyte they contain, are the PEM, which has a solid polymer electrolyte; solid oxide, with a ceramic electrolyte; and phosphoric acid and alkaline, both with liquid electrolytes. Each is at a different stage of development (see Table 5-1).

PEM fuel cells appear to be the most promising for use in vehicles. The distinguishing feature of PEM fuel cells is the use of a thin plastic membrane that looks like "Saran" wrap for the electrolyte. Using a solid material for the electrolyte allows the PEM fuel cells to be more compact and to operate at a cooler temperature than those with liquid electrolytes.

The emergence of PEM technology as the preferred fuel cell candidate

TABLE 5-1
Characteristics of Fuel Cells, 1994

	Status	Power Density (kw/liter)	Temperature (°C)	Contaminated by
Alkaline	Spacecraft	0.16	150–250	CO, CO_2
Phosphoric acid	Commercial	0.1–1.5	65–220	CO
PEM	Demonstration	0.1–1.5	25–120	CO
Solid oxide	Laboratory	1–4	700–1,000	—

Source: Updated from Mark A. DeLuchi, "Hydrogen Fuel-Cell Vehicles" (Report 92-14, University of California, Davis, Institute of Transportation Studies, September 1992).

for vehicles is the result of major improvements in performance achieved in the early 1990s by Ballard Power Systems, as well as greater awareness of this technology's potentially low cost. Low cost is possible because the PEM materials—graphite for the plates and thin plastic sheets for the electrolyte—are inherently inexpensive. Platinum is the only expensive material needed; it is a catalyst placed on the plates to help the electrons split from the hydrogen molecules. But even this cost is dwindling as researchers learn to use less platinum.

Other advantages of the PEM fuel cell are low operating temperature, rapid response to shifting power demands, and the potential for downsizing.[3]

The principal disadvantage of the PEM fuel cell is that, unlike its phosphoric acid and solid oxide counterparts, it is sensitive to carbon monoxide, a by-product in the reforming of methanol. As a result, the PEM fuel cell is less well suited to the use of methanol. General Motors and Mercedes Benz are both adapting PEM fuel cells to operate on methanol—apparently with success.

Though PEM fuel cells are now the leading contender, one should be cautious about premature predictions. Fuel cell technology is developing rapidly. Only a few years ago in their mammoth treatise on fuel cells, A. J. Appleby and F. R. Foulkes concluded that "the alkaline hydrogen-air fuel cell is the most promising system for use as a power source for electric vehicles."[4] They did not even rate PEM fuel cells as an attractive option. It remains possible that an even better fuel cell technology than PEM will emerge.

As for the alkaline fuel cell, it is now considered less promising for vehicle applications, despite considerable experimentation and development at NASA. It offers quick start-up, but is intolerant of carbon dioxide and requires pure hydrogen, which is prohibitively expensive.

Phosphoric-acid fuel cells have also received considerable attention. In the mid-1980s the U.S. Department of Energy made the decision to power its three buses with them, mostly because these fuel cells were more proven and available than PEM and other candidates. Phosphoric-acid fuel cells have the advantage of being more tolerant of carbon monoxide, but they are bulkier than PEM cells and take more than fifteen minutes to warm up. The size and delayed start-up might be acceptable for locomotives and buses, but not for private cars.

The solid-oxide fuel cell is the least developed. Its principal advantage is in combining the functions of reformer and fuel cell. By reforming methanol or methane directly within the cell, the solid-oxide version saves space and possibly cost. It is also less sensitive to the type of fuel used. Its principal disadvantage is a high operating temperature, which creates several possible warm-up and safety problems. Further development might diminish these obstacles. As of late 1994, a full solid-oxide fuel cell system still had not been built or tested. The cost of this technology is highly uncertain.

The future of fuel cell vehicles depends not on technical break-throughs but streamlined engineering. One challenge is to design and integrate peak-power devices into fuel cell systems. One fuel cell expert suggests that enough hydrogen could collect on the plates to provide surge power for up to ten seconds, possibly eliminating the need for peak-power devices altogether.

A second challenge is to design the fuel cell, peak-power device, motor, electronics, and fuel-storage system to fit into as small a space as possible without compromising safety. For example, high-voltage components must be completely isolated. Modular components and the electronic connections between them afford designers and systems engineers some flexibility in separating and arranging the parts.

Another challenge, if methanol is used, is to design reformers that take much less time to warm up and that can respond to rapid changes in power demand, for instance, hard accelerations. An alternative would be to design internal-reforming solid-oxide fuel cells.

Many other more specific technical challenges face each type of fuel cell. For PEM cells these include improving the performance and reducing the cost of the electrolyte membrane without compromising its mechanical properties or making it susceptible to impurities in the gas stream; finding a simple and effective way to keep the membrane moist, while at the same time not allowing product water to build up at the cathode; reducing the size and energy consumption of the air-compression system; and reducing the weight, bulk, and manufacturing cost of the stack plates and assembly.[5]

Hydrogen as a Fuel

Fuel-cell vehicles have many advantages, including zero emissions, quiet operation, long range, and unparalleled energy efficiency. Perhaps most compelling of all, though, is that they might realize the elusive dream of solar hydrogen. Fuel cells run on methanol, natural gas, or petroleum would dramatically reduce pollution and cut greenhouse gas emissions and energy consumption at least by half. But solar hydrogen would push the fuel cell one giant step further—to a future that came close to being environmentally benign. No other transportation technology or fuel holds the same promise.

Solar Hydrogen

Solar hydrogen is hydrogen made by electrolytically splitting water using solar cells. The technology is still expensive. However, Joan Ogden and Robert Williams of Princeton University, along with others, have argued that rapid advances in solar technology make inexpensive solar electricity a possibility.[6] They predict that the cost of solar hydrogen in the early years of the next century, including all distribution and retail markups but not taxes, could be between $2.25 and $3.25 per gasoline-equivalent gallon—about two to three times the future price of gasoline.[7]

What makes solar hydrogen potentially feasible is that fuel cells are more than twice as energy efficient as ICEs. The relatively high cost of solar hydrogen is, from the vehicle user's perspective, more than halved, bringing fuel costs within a commercially viable range.

Currently, the least expensive and most common way to make hydrogen is by reforming natural gas. The cost is about $0.70 per gasoline-equivalent gallon, slightly more than gasoline. (At present, most natural-gas-derived hydrogen is used in chemical plants and oil refineries.) Until the price of cleanly produced hydrogen drops and environmental factors play a larger role in selecting fuels, natural gas reforming will remain the preferred method for producing hydrogen.

The most environmentally benign method for producing hydrogen is to split water by means of clean electricity; and the least expensive form of clean electricity to split water is electrolysis.[8] In electrolysis, an electrolyte is added to water to make it more conductive; then an electric current is applied to the solution, producing hydrogen gas. The electric current can come from any source; the most benign source, and not necessarily the most costly, is photovoltaic solar cells. Solar hydrogen, then, is simply hydrogen made by electrolytically splitting water with solar cells.

The ideal place for producing solar hydrogen is in sunny locations such as the American southwest. If all the cars and trucks in the United States were fueled with hydrogen in 2010, only 1 percent of the deserts in this land (equivalent to 0.1 percent of the total area of the country) would be needed to produce the hydrogen.[9]

Solar hydrogen facilities are sparing of natural resources. The typical site would require only about an inch of annual rainfall over the area of the solar cells, and no fossil fuels or chemicals.[10] It would contribute no toxic waste, generate no air pollution or greenhouse gas, and require a minimal on-site staff. In other words, the facility would have a negligible effect on the local ecology. Solar hydrogen would be much more environmentally benign than biomass, coal, natural gas, and all other large-scale energy systems.

Biomass-Derived Hydrogen

A complement and possible transition to solar hydrogen is hydrogen made from biomass. As indicated in Chapter 2, the cheapest and most available biomass crops are trees and grass. Like wood pulp and Christmas trees today, they could be grown on high-yield plantations and harvested regularly—trees every few years, grass several times per year. The biomass would be converted to methanol or hydrogen at relatively small processing plants located on or near plantations.

The chief attractions of biomass are in the emissions and the cost. Biomass offers much lower overall greenhouse gas emissions than fossil fuel–based energy systems, and it is cheaper than solar hydrogen. But biomass production is likely to be limited. The cost of production increases sharply with larger volume, both at individual production sites as well as on a national level. Biomass crops must be grown in relatively compact areas. The cost of transporting raw cellulosic biomass to a central processing plant is high; if the land surrounding a particular production plant is not fertile, or if not enough of it is devoted to the biomass crop, costs escalate sharply.[11]

Perhaps more important, large-scale biomass production puts a strain on local ecological systems. Intensive monoculture farming reduces biodiversity and erodes soil; soil runoff pollutes water, and small, decentralized production plants pollute air. These environmental drawbacks, along with limited opportunity for production, will likely rule out biomass as the dominant source of chemical energy for motor vehicles.

Hydrogen Transport

From remote sunny areas or biomass plantations, hydrogen would be transported to urban centers by pipeline. Two options are possible: build new pipelines or use existing natural gas pipelines. New pipelines would not pose technical problems. Despite a high initial investment, the cost per unit of hydrogen in an efficient system would be small. The only real problem with new pipeline construction may be political opposition in developed urban areas.

Transport of hydrogen in existing natural gas pipelines is appealing but problematic, principally because hydrogen embrittles certain kinds of steel pipe, causing cracks and leaks. An unresolved question is whether chemical compounds could be added to hydrogen to inhibit this process.[12] Recent research suggests that safe, effective, and inexpensive inhibitors do exist, but this has not been fully verified in extended tests.

Hydrogen Storage

A greater technical and economic challenge of hydrogen is on-board storage. Hydrogen, the lightest element on Earth, is difficult to store compactly. The following storage techniques, at various stages of development, hold some promise: compressing the gas under high pressure; liquefying it at very cold temperatures (cryogenic liquefaction); and embedding it in metal or liquid hydrides, sponge iron, or activated carbon. Each technique would be significantly improved—in terms of cost, bulk, size, and complexity—with further research and development.[13]

The most feasible method for the near term appears to be high-pressure storage. Compressed gas is rather bulky, even at very high pressure, but it is light and quick to refuel, and the technology is fairly simple. Cryogenic liquefaction systems are light and compact, but refueling would be expensive and probably perceived as unsafe because of very low temperatures. Current systems that embed hydrogen in metal or liquid hydride structures are bulky, heavy, and complex.

Carbon adsorption and sponge-iron technology, though furthest from commercialization, show considerable promise. In carbon adsorption, hydrogen is forced into a pressurized, refrigerated tank where it adheres to activated carbon, a highly porous material. The sponge-iron technique was once used to produce hydrogen commercially. Hydrogen is generated on board by using steam to oxidize powdered iron stored in a tank. The iron or the entire tank can be replaced in minutes. Sponge-iron tanks may prove to be the least expensive and bulky system, but probably heavier and more complex than the best of the other options.

If compressed hydrogen proves unacceptable, methanol will by default become the preferred hydrogen carrier in the near term. Methanol is an attractive carrier because it can be reformed into hydrogen relatively easily and inexpensively. As noted earlier, the other near-term fuel choice, reforming natural gas, is not feasible on board vehicles. Reformers are so large and bulky that they would need to be used off board, with hydrogen boarded and stored in vehicles.

Methanol itself has drawbacks. While it works well in phosphoric-acid fuel cells, those fuel cells are not well suited to cars and most trucks. Methanol use in PEM fuel cells may be problematic. The on-board methanol reformer, though simpler and much smaller than a natural-gas reformer, would still be bulky and expensive. More importantly, it would generate impure hydrogen, increasing the complexity and cost of PEM fuel cell designs.

Safety Considerations

For successful marketing, fuel cell vehicles must prove safe. Hydrogen is perceived not to be so, a legacy of the flaming crash of the *Hindenburg* dirigible in 1937. That disaster, though dramatic, unfairly tarnished hydrogen's reputation. (In fact, many of the thirty-five deaths—of ninety-seven passengers in all—were caused by people jumping and by burns from the *diesel* fire. Most who waited for the burning ship to land walked away unharmed.)[14] Until policymakers and the public are convinced that hydrogen is no more dangerous than the petroleum fuels to which they are accustomed, fuel cell vehicles will be unacceptable.

In the technical community, it is widely agreed that hydrogen is not inherently more dangerous than gasoline. The hazards of hydrogen are different but not necessarily greater than those presented by current petroleum fuels. It is worth remembering that over a hundred years ago quite similar objections were voiced against gasoline. After a few years, the apparently tolerable safety record of gasoline dispelled the most serious concerns.

The Cost of Fuel Cell Vehicles

Cost is always the key question in commercialization. It is particularly difficult to project for fuel cell vehicles because the technology is so young, but present trends are highly encouraging. In cost and performance, fuel cells are improving exponentially. Researchers from General Motors recently predicted that the cost would drop from as much as $10,000 per kilowatt in 1992 to as little as $20 in the future, at which level a fuel cell vehicle would cost the same or less than a gasoline-powered vehicle.[15] Even with more conservative projections, though, because of their lower operating costs, fuel cell vehicles could have considerably lower life-cycle costs than comparable gasoline or battery-powered vehicles.

The cost of fuel cell vehicles is sensitive to the type of fuel used. As shown in Table 5-2, the life-cycle costs of fuel cell cars powered by solar hydrogen may be somewhat greater than those of comparable gasoline cars, whereas vehicles powered by methanol made from biomass would be considerably less expensive over the life cycle. Vehicles run on hydrogen made from natural gas would be the cheapest. These findings are not definitive. They should be treated not as projections but rather as a

TABLE 5-2

Long-Term Life-Cycle Costs of Mid-Size Gasoline-Powered
and Fuel-Cell Cars (cents per mile)

	Gasoline (400-mile range)	Fuel Cell, Solar Hydrogen (250-mile range)	Fuel Cell, Biomass Methanol (350-mile range)
Vehicle (excluding battery, fuel cell, hydrogen storage)	17.4	14.0	13.9
Battery, tray, auxiliaries	—	2.0	2.0
Fuel (including taxes)[a]	5.8	6.2	4.4
Fuel-storage system	—	1.5	—
Fuel-cell system	—	2.3	2.8
Maintenance and repair	3.4	2.4	2.5
Other[b]	2.0	1.0	1.0
TOTAL	28.6	29.4	26.7

[a] Taxes are identical per mile for all fuels.

[b] Includes inspection and maintenance fees, oil for the gasoline car, insurance, registration, and tires.

scenario of what is plausible under mass production and with continued improvement in key technologies, including solar electricity. No major technological breakthroughs are assumed.

The most important cost relationships and assumptions of this analysis are consistent with those presented in Chapter 3 for battery-powered vehicles, and are based on the same Delucchi cost model. One difference is that a mid-size vehicle is used here instead of a subcompact, because fuel cells are suitable for larger vehicles.

Why are life-cycle costs for fuel cell vehicles relatively attractive? First, the electric power train (excluding the battery and fuel cell) would probably cost about 10 percent less than the ICE system it replaced, and last about 15 percent longer. Second, fuel cell vehicles would likely be more than two times as energy efficient as comparable gasoline cars, bringing the per-mile cost of solar hydrogen down almost to that of gasoline cars, and even lower if the hydrogen were made from natural gas or methanol. (Hydrogen made from water is assumed here to cost $3 per gasoline-equivalent gallon, compared with $1.20 for gasoline, excluding taxes.) Third, maintenance of the electric drive train would probably be significantly cheaper.

If technologies do not improve as expected, fuel cells will be less competitive than indicated in Table 5-2. If, on the other hand, unanticipated breakthroughs do occur, or very optimistic projections such as General Motors' are realized, then the cost of fuel cell vehicles would be even lower than indicated. Robert Williams of Princeton University, for instance, calculates that the life-cycle cost of PEM fuel cell vehicles operating on hydrogen made from natural gas, methanol, biomass, or water would be 5 to 12 percent less than that of gasoline vehicles.[16]

Several conclusions emerge from the cost analysis in Table 5-2 and earlier analyses.

1. Hydrogen fuel cell vehicles will probably have a lower life-cycle cost per mile than hydrogen ICE vehicles, primarily because electric drives are more efficient and less costly than ICE systems. An additional bonus is lower emissions, especially the elimination of troublesome nitrogen oxides. Hydrogen use in fuel cells promises to be far superior to hydrogen use in ICEs, though hydrogen fuel cells will not be available as soon.

2. Hydrogen fuel cell vehicles will probably have lower life-cycle cost than battery-powered electric vehicles, except perhaps in small vehicles with very short range. Fuel cells are better suited to longer-range vehicles because as energy demands rise, the weight and cost of fuel cell systems, even with hydrogen storage, go up much more slowly than the weight and cost of batteries. A large vehicle with extended range requires more batteries, which adds considerable weight and thus requires still more batteries and more support in a continuous upward spiral. On larger cars and cars with extended range, batteries could account for more than one-third of the weight of a vehicle. Fuel cell vehicles are like gasoline cars: range is gained by simply expanding the size of the fuel tank. While hydrogen storage is expensive, it is much cheaper than expanding the size of batteries.

3. Life-cycle competitiveness does not depend on large reductions in the cost of the fuel cell itself. Other factors, such as vehicle life and maintenance costs, can be just as important.

4. Methanol fuel cell vehicles will probably have a lower life-cycle cost than hydrogen fuel cell vehicles—when the methanol and hydrogen are made from the same source. The fuel cell system is more expensive for vehicles fueled by methanol than for hydrogen-fueled vehicles because an on-board reformer is needed to convert methanol into hydrogen. The reformer also reduces fuel

efficiency by about 20 percent. This disadvantage of methanol is offset by the far greater ease and cost of storing it on board.

5. Fuel cell vehicles run on solar hydrogen will be more expensive than other options considered in this book, but will cause much less environmental damage.

Environmental Advantages

A hydrogen-powered PEM fuel cell is a true ZEV. Water is the only effluent, and the vehicle emits no air pollutants. Methanol fuel cell vehicles would produce trace amounts of nitrogen oxide and carbon monoxide from the methanol reformer, and small amounts of evaporated methanol from the fuel supply and storage systems (there have been no driving tests, so exact amounts are unknown). Emissions from fuel cell vehicles operating on gasoline or other petroleum fuels would be somewhat higher still, not from the fuel cell but from the reformer, the fuel tank, and upstream sources.

In addition to running cleaner than ICE vehicles, fuel cell vehicles would produce less noise pollution. Some fuel cell auxiliaries—pumps, blowers, and compressors—would be audible, and the electric motor would hum and whine, but because the electrochemical reactions in a fuel cell are silent, the net effect would be much less noise, especially at lower speeds.

Making the Transition to Fuel Cells

The easiest way to change large complex systems is with a stream of small incremental changes. The easiest way to introduce fuel cells would be to use petroleum fuels, reformed into hydrogen on board the vehicle. This strategy has the advantages of sharply reducing energy consumption, greenhouse gases, and other air pollutants, yet retaining the existing infrastructure of petroleum refining, storage, and transport. It would also undermine opposition from the oil industry. At least one major oil company, already thinking along this line, is funding the development of reformers for converting petroleum fuel into hydrogen. Once petroleum-powered fuel cell vehicles appear in the showroom and on the road, the transition to more benign fuels could be accomplished gradually, awaiting improvements in on-board hydrogen storage and following the creation of a new fuel-supply infrastructure.

If small, inexpensive reformers for petroleum fuels prove infeasible, then the next easiest strategy would be to use natural gas or methanol for fuel.

Methanol is relatively cheap and, being a liquid, easy to store and transport. The oil industry could adapt its extensive petroleum distribution system to methanol, though at some cost and with some inconvenience.[17] The cost of methanol distribution would be high initially, reflecting dependence on expensive truck and rail transport, until sales volumes were great enough to justify construction of new pipelines (or conversion of existing petroleum pipelines).

The fuel cell technologies best suited to methanol are phosphoric acid and solid oxide, though PEM fuel cells are likely to prove satisfactory. Methanol, made from natural gas, should be plentiful through much of the next century at relatively low cost. While methanol currently costs about 10 to 30 percent more than gasoline on an energy-equivalent basis, this premium should shrink over time as petroleum becomes scarcer. Methanol could also be made from cellulosic biomass at a somewhat higher cost.

Alternatively, natural gas could be the source of hydrogen. This option is attractive primarily because natural gas transmission and distribution systems are well established and would require no modification, a plus compared to methanol. Also, natural gas is relatively inexpensive, even with reforming.

Natural gas resources are abundant in many countries, including the United States. Because fuel cells use less than half as much energy as ICEs, total natural gas consumption would be modest. As fuel cell vehicles proliferate, there would be plenty of time to switch to solar hydrogen or biomass methanol.

The natural gas reforming strategy is not problem-free, though. As mentioned earlier, because on-board reforming of natural gas does not appear practical, natural gas would probably have to be reformed off board and hydrogen stored on board. However, on-board hydrogen storage is difficult, and local residents might oppose the establishment of natural-gas reforming stations in their neighborhoods.

It would seem, mostly for economic and political reasons, that an incrementalist approach is sensible, whether with hydrogen, methanol, or petroleum. But credible arguments can also be made for moving directly to solar hydrogen or biomass methanol. The disadvantage of this more aggressive strategy is that the transition would undoubtedly begin slowly, mostly because of the higher cost of hydrogen and the time it would take to create a fuel distribution system for hydrogen. However, there would be benefits. First, this strategy recognizes the obvious—that fossil fuels will eventually be phased out anyway, probably during the next century. This is because they are finite energy sources and the principal contributor of greenhouse gases. Second, hydrogen made by splitting water with solar energy is far superior environmentally to natural gas–based fuels. Third, even an incremental transition would be difficult. We cannot realistically expect to summon sufficient political will to arrange a major industrial transformation every twenty years. If the ultimate goal is known, perhaps it would be better to move slowly but directly toward that goal, reducing the huge uncertainty associated with an incrementalist strategy. With an incrementalist strategy, uncertainty arises from companies betting that demand for methanol (and perhaps liquefied natural gas imports) would or would not remain strong long enough to gain a profit from the multibillion dollar investments that would have to be made in new facilities.

The incrementalist strategy is clearly the safer and easier one. But if energy and environmental crises intervene—perhaps as a result of turmoil in the Persian Gulf, or new evidence of climate change—then a more aggressive jump to biomass methanol and solar hydrogen should be seriously considered.

- - - - - - - - - - - - - -

If technology continues to develop at a moderately rapid clip, fuel cells will prove to be a significant advance over both internal combustion and battery-powered vehicles. They would be cleaner and more efficient than internal-combustion vehicles and cheaper and better performing than battery-powered electric vehicles.

However, the research and development challenges in this area are not trivial, nor are the design strategies straightforward. For each engineering task, there are various technology and design routes to consider—at least three different kinds of fuel cell potentially suitable for highway vehicles, at least four different ways to supply peak power (batteries, ultracapacitors, flywheels, or the fuel cell itself), and many ways to store hydrogen. While multiple options make development and design complex, they also increase chances for the eventual emergence of a successful fuel cell vehicle. More aggressive support of fuel cell research and development by government and industry can accelerate the timing of this process. It is conceivable that fuel cell buses could be commercially available by 2003, and fuel cell cars a few years later.

Ultimately, marketability will be the yardstick of success. To probe how consumers would use and react to fuel cell vehicles, and to help guide choices with respect to range, power, ease of refueling, reliability, safety, and cost, numerous experimental vehicles should be built and tested. The purpose of these should not be to display a purportedly finished technology but rather to glean information from consumers for feedback into basic research and development. Fuel cell vehicle technology is already far enough along to begin building experimental models.

Well-designed fuel cell vehicles could be the primary component of a strategy for reducing dependence on imported oil, mitigating global warming, and improving urban air quality. Fuel cell technology appears to be the most attractive alternative to petroleum fuels and internal combustion engines to emerge in many years, perhaps ever.

CHAPTER 6

Hybrid Vehicles: Always Second Best?

In the broadest sense, a hybrid is a vehicle with two different propulsion systems. Here we will look at the hybrid more narrowly, as a combination of an electric motor and an internal combustion engine (ICE) in a single vehicle. The broader definition would consider the use of fuel cells as substitutes for the ICE, but the subject of fuel cells has already been addressed at length.

The main idea behind hybrid vehicles, and their primary attraction, is that they extend range beyond what batteries by themselves can provide. They also have the potential to produce less pollution and consume less energy than ICE vehicles. The disadvantage of hybrid vehicles is that by combining two propulsion systems and their associated energy storage units, they are inherently more complex than both pure battery-powered electric and pure internal-combustion vehicles.

Despite their advantages, hybrid vehicles are a middling choice whose future is far from assured. They are likely to be a victim of regulatory circumstance in the near term, pushed aside by improvements in gasoline emissions and by the ZEV mandate. In the longer term, they will be squeezed between improving batteries and emerging fuel cells. Hybrid vehicles have considerable promise and could play a central role in creating a more sustainable transportation system. But only if fuel cells and advanced batteries falter.

Hybrid Vehicle Development

First built in the early 1900s by inventors tinkering with combinations of the electric motor and the gasoline engine, hybrid vehicles were dropped when gasoline-fueled vehicles became more reliable and easier to start, and gasoline fuel more readily available. Research and development of hybrid vehicles was revived by concern about oil dependency in the 1970s and about air pollution in the late 1980s.

The principal financial support for hybrids has come from the U.S. Department of Energy. Some automobile companies have also invested in the technology. But by early 1994, Volkswagen was the only automaker seriously considering marketing a hybrid vehicle—only in Europe, though, not in the United States.

A number of hybrid vehicles have been built and tested since 1980.[1] Some of these vehicles have impressed analysts with their performance and low levels of exhaust and petroleum consumption. Interest in hybrid vehicles jumped in late 1993 with the announcement of funding for two major collaborations. The Department of Energy signed a five-year, $138 million development agreement with General Motors and a $122 million agreement with Ford to design and build preproduction hybrid prototypes that could be marketed in less than 10 years. Costs are being split equally between the government and the companies.

The General Motors hybrid vehicle will have a ceramic gas-turbine engine as the primary power source, with an electric motor and batteries for peak power. The Ford program, not yet formalized as this book goes to press, hasn't announced what sort of hybrid it will build. But it is clear that the primary objective in both programs is to improve the fuel economy of full-size cars by at least 50 percent over that of current gasoline-powered cars, and to reduce emissions to California's "ultra-low" level.

Hybrid development has been sluggish partly because of uncertain market demand and confusion over design goals. By rearranging and altering the principal components of a hybrid vehicle—the electric motor, ICE, transmission, and battery—one can meet differing design objectives. Should components be arranged principally to minimize energy consumption, or emissions? To reduce emissions to zero in selected areas such as polluted downtowns? Or to satisfy customers by extending driving ranges at minimal cost and with minimal disruption to the status quo?

Even if consumer demand were clear and design goals agreed upon, rapid improvements in electric drive-train technology require a rethinking of the optimal size of the motor, engine, and other components.[2] Electric drivelines, excluding batteries, are now considerably smaller and lighter than engines and transmissions of the same power output, and probably less expensive when mass-produced, creating design options not available as recently as the late 1980s.

The Family of Hybrid Vehicle Designs

Hybrid vehicles can fill the spectrum of vehicle options from pure battery-powered electric vehicles to conventional ICE vehicles. At one end of the spectrum, closest to pure electric vehicles, is the so-called *range-extended* battery hybrid. It would utilize a 50 to 100 kilowatt (kw) electric motor and batteries to provide good acceleration and performance and all-electric range of about 50 miles. It would operate in an all-electric ZEV mode most of the time, carrying a small 5 to 10 kw engine to extend range an additional 50 miles, but not at sustained highway speeds. As batteries improve, the all-electric ZEV range could be extended or, alternatively, the battery pack made smaller.

A second option along the spectrum is the *dual-mode* hybrid. This vehicle would also have a ZEV capability of 50 miles or so, but with a much larger 25 to 40 kw engine for driving longer distances or at higher speed. When this hybrid operates in an all-electric ZEV mode, it has zero tailpipe emissions and burns no petroleum.

Unlike the range-extended battery hybrid, the dual-mode hybrid with its larger engine could substitute for a conventional gasoline-powered vehicle. However, when the hybrid is operated in the non-ZEV mode, its emissions and petroleum consumption would be substantial.

Approaching the conventional ICE vehicle is the *fueled engine–electric* hybrid. All the energy to power this hybrid would come from gasoline or some other chemical fuel stored on board. The engine would act as a generator, creating electricity for the electric motor, with excess electricity stored in an on-board battery, flywheel, or ultracapacitor. The engine would operate intermittently, at a near-constant speed when on. This is the hybrid design embodied in the supercar concept promoted by Amory Lovins.[3] Emission levels and gasoline consumption would be much lower for this hybrid than for a comparable ICE car, but, unable to run for extended periods in an all-electric mode, it would not qualify as a ZEV even with the most liberal rules. And because its only source of energy is gasoline (or some other chemical fuel), the fueled engine-electric hybrid would consume much more petroleum (or natural gas or alcohol) than other hybrids.

Hybrid Technology

Hybrids are typically characterized by their driveline configuration as either parallel or series vehicles. While not useful for analyzing policy, regulatory, or marketing issues, this distinction is helpful as a framework for describing hybrid technology.

Parallel Configurations

In a parallel hybrid, both the engine and electric motor can drive the wheels. The output of the electric motor is added as needed directly to the power output of the engine; in other words, the electric motor assists the engine, in a reversal of the roles played in a series arrangement (see Figure 6-1). Consequently, in a parallel hybrid the electric motor can be smaller than the engine, playing a correspondingly smaller role in the vehicle's operation. The electric motor in a parallel hybrid will generally be smaller than that in a series hybrid, depending on the design. Of the three types of hybrids just described, parallel drivelines are best suited to the dual-mode hybrid.

Because high-power electric-drive systems would have been expensive, bulky, and complex in the 1980s, electric drives until recently were designed to be small, performing as a complement to a gasoline engine rather than as the primary power source. The parallel design suited this arrangement. The typical approach with early hybrids was to downsize the gasoline engine slightly and to add batteries and a low-power electric motor for power assistance in hard accelerations and hill climbing

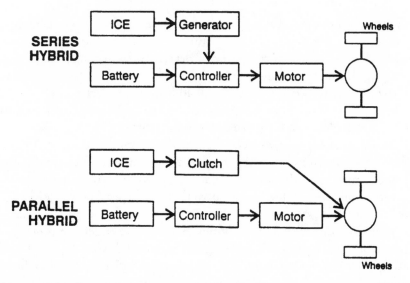

Figure 6-1 SCHEMATIC DRAWING OF SERIES AND PARALLEL HYBRID CONFIGURATIONS.

(similar to the role peak-power devices such as ultracapacitors are likely to play in future electric and hybrid vehicles).

Initially, parallel drivelines had the added advantage of greater energy efficiency. In a series hybrid, the process of converting the chemical energy of gasoline (or other fuel) to mechanical, then electrical energy, and then back to mechanical energy to drive the vehicle is less efficient than converting chemical energy directly to mechanical energy, as in a parallel hybrid. This advantage is diminishing, however, as more advanced electronics allow a larger role for the electric component in the propulsion system and more efficient management of the various driveline components. (These factors also enhance parallel hybrids, but less so.)

Overall, parallel hybrid vehicles will likely be more complex and more expensive than series hybrids. In parallel hybrids, the engine must be turned off and on frequently, additional mechanical components are needed to link the engine and electric motor, and complex control algorithms blend engine and motor output. Driveline control is more complex because there are circumstances in which either the electric motor or the engine powers the vehicle, and other circumstances in which the two operate together.

Series Configurations

A series hybrid uses the fuel-burning engine to generate electricity and the electric motor to drive the wheels. All of the mechanical energy from the ICE is converted on board to electricity and fed directly to the electric motor or, if not needed at the moment, to the battery or ultracapacitor for storage. There is no clutch or transmission in the engine system because all mechanical energy is converted to electricity for driving the wheels. A series configuration is more logical for the range-extended hybrid because the engine is essentially a small add-on to the electric drive system. This configuration would also be used for the fueled engine-electric hybrid, and could be used for the dual-mode hybrid.

In the series hybrid, especially the fueled engine-electric version, the ICE would run for extended periods at nearly constant speed and power, being turned off only when the battery approached full charge. Power surges would be provided by the battery or other peak-power device, not the engine.

A major issue with all battery-powered electric vehicles, especially series hybrids, is battery life. Shifting the responsibility for rapidly fluctuating power from the engine to the electric motor sharply reduces the engine's energy consumption and emissions, but at the cost of shortening battery life. Today's lead-acid batteries, along with most batteries soon to be commercially available, are designed for daily recharge, not tens or hundreds of charging and discharging cycles each day. Series hybrids will require new batteries designed for high power density and to withstand a large number of shallow, rapid charge/discharge cycles.

Because series designs will benefit more than parallel designs from expected rapid advances in electric-drive technology and continuing improvements in battery technology, it is likely that series hybrids will slowly supplant parallel hybrids.

Complexity: Less of a Problem

Although it may be necessary to refuel two different energy systems, driving a hybrid vehicle should be no different than driving a conventional engine-powered vehicle. In both cases, the driver will control speed with accelerator and brake pedals. However, because the motor and engine need to be coordinated, managing the driveline to respond to driver demand will be much more complex for the hybrid than for an all-gasoline or all-electric vehicle. Depending on the driveline configuration, at any particular moment either the motor or engine or both may have to kick in to meet the driver's power demand.

The control strategy for the hybrid will determine when to turn the motor and engine off and on and how much power each will provide under different driving conditions. Sensor instruments and an on-board computer will determine the operating status of the driveline components. Such computerized sophistication was virtually unimaginable until recently, but no longer. Computerized controls are now used in conventional cars to smooth operation of the engine and automatic transmission. Future control of the hybrid driveline will be a more extensive application of present microprocessor technology, making complexity less of a problem.[4]

Flywheels and Ultracapacitors for Peak Power

Parallel-configured hybrids from the 1970s and 1980s relied on batteries for peak power. Future ICE-electric hybrids, as well as pure electric designs (including fuel cells), will have the option of meeting peak power needs more efficiently and cheaply with flywheels or with high-energy density capacitors, often referred to as ultracapacitors.[5]

These peak-power devices will power the vehicle during acceleration and accept regenerative braking energy during deceleration. Their energy storage capacity must be great enough to prevent excessive discharging during acceleration. For most vehicle designs, peak power devices would be expected to store less than 1 kilowatt-hour (kwh) of energy, compared with 20 to 40 kwh for the battery in a pure electric vehicle. The chief attractions of flywheels and ultracapacitors are an ability to withstand many more charge/discharge cycles than batteries, and to transfer and receive energy more rapidly. These devices also reduce and smooth demand on primary batteries, thereby prolonging their life and making possible smaller engines and battery packs. They are the critical enabling technologies for hybrid vehicles. A disadvantage is that peak power units add to the complexity of hybrid designs.

Flywheels, or electromechanical batteries, store energy in a lightweight, carbon composite rotor that spins at over 100,000 revolutions per minute. The flywheel system includes a rotor, hub, shaft, bearings, containment vessel, motor, and electronics.

Flywheel technology received extensive and favorable attention during the 1970s, when it was developed, and again in the 1990s, when advances were made in composite materials and electronics. Several national laboratories and corporations in the United States are developing flywheels.[6] Remaining technical obstacles include the general complexity of the system, the need for magnetic bearings and a lightweight containment vessel suitable for use in vehicles, development of a high-power motor and associated electronics, and the high cost of carbon composites. If all these challenges can be overcome, the flywheel will be an attractive means of storing energy on board a hybrid vehicle, especially for peak power.

Ultracapacitors are a newer, simpler, safer, and potentially cheaper peak-power device.[7] A capacitor is one of the oldest ways of storing electricity. It allows a charge to build up between two conductive plates separated by an insulator. The plates are made of a porous, conductive material impregnated with an electrolyte. Electricity is stored in the extensive surface area of the plates.

Low-power ultracapacitors are commercially available today in computers and small appliances such as VCRs, telephones, and alarms, but they have energy densities of less than 1 watt-hour per kilogram (wh/kg). High-power ultracapacitors for hybrids and pure electric vehicles will need an energy density of at least 5 wh/kg.

Efforts are under way at national laboratories and several companies to improve the energy density, life, and cost of capacitors with porous carbon and ceramic mixed-metal oxides. Progress has been rapid. As of early 1994, prototype cells using these materials had attained energy densities ranging from 2 to 10 wh/kg, and densities as high as 30 wh/kg appear to be possible. Laboratory tests have also attained efficiencies of over 90 percent and lives exceeding 500,000 cycles.

Advanced prototypes are expected to be ready for testing by late 1995 and for the commercial electric and hybrid vehicle market shortly thereafter. By the early twenty-first century, it is possible that the battery pack in a fueled engine–electric hybrid could be replaced with ultracapacitors for only about $500 to $1,000. Ultracapacitors would be several times cheaper than the batteries they replaced, and they would last much longer.

Prospects for Hybrids

Rapidly developing electric components are expanding the opportunities for hybrids far beyond what was possible only a few years ago. It will soon be possible to design both series- and parallel-hybrid drivelines so that vehicles can be operated in a ZEV mode for most city driving, with performance comparable to or better than that of conventional gasoline-powered vehicles. It will also be possible to build a series hybrid that runs exclusively on gasoline or some other fuel, minimizing the need for a recharging infrastructure and still significantly reducing gasoline consumption.

Considerable research remains, however, before an advanced, marketable hybrid is built. Key components are already commercially available or in advanced stages of development. The challenge is to integrate them in a unit with extended range, suitable power, good performance, and low emissions at an acceptable cost. The effort to produce hybrids does not appear beyond the capability of the automobile industry, but the longer investment lags, the less likely it is that hybrids will play a large role in the automobile market.

Environmental and Energy Considerations

Shifting the function of peak power supply away from engines presents the possibility for major improvements in emissions and energy consumption. Why? First, the engine can be downsized by three-quarters or more to provide only the vehicle's average, not its peak, power. The smaller engine will be more energy efficient than the large gasoline-powered engine when it is operating at low power, the condition typical of conventional ICE car driving. Small engines are already used in motorcycles, boats, and various industry applications, but they would need to be re-engineered for hybrid applications where minimum fuel consumption and low emissions are critical.

Second, it is far easier to reduce emissions when an engine slowly varies speed and power. Indeed, recent evidence indicates that most pollution from gasoline engines is emitted during hard accelerations and decelerations.[8]

Third, constant or near-constant engine speed greatly enhances the feasibility of using inherently more energy-efficient ICEs, especially gas turbines and two-stroke engines.[9] This advantage could be very important. Gas turbines are more efficient at full-load, steady-state operation, which is how they are used in jet aircraft. They are not well suited to the wildly varying engine speeds of stop-and-go traffic. In a series hybrid, the turbine can run at optimal speed because it is attached not to the wheels but rather to an alternator that generates electricity. Volvo's innovative series-hybrid prototype, unveiled in 1993, has a gas-turbine engine. General Motors' hybrid development program also features a gas-turbine engine.

Unfortunately, no credible analyses of hybrid emissions and energy consumption have been done. Previous studies were plagued by methodological and data problems, including wide variation in possible hybrid designs; uncertainty regarding the percentage of time a hybrid's ICE would actually run; and a lack of tests for accurately measuring hybrid emissions and fuel economy. Analyses have failed to account for actual driving conditions; to place hybrid, pure electric, and pure ICE vehicles on the same footing; and to relate hybrid technology to policy considerations.

It is beyond the scope of this book to resolve these analytical shortcomings, but certain analytical information is useful for policymakers and regulators. The analyses of energy use and emissions in Tables 6-1 and 6-2 were prepared under the direction of A.F. Burke. Results presented in these tables are not definitive because of the data problems cited above, but are indicative of relative differences between vehicle types. The analyses presented here rely on models developed at the Idaho National Engineering Laboratory.[10]

We ran simulations for several hybrid and electric vehicle designs.[11] All vehicle attributes except those related to the driveline were held constant. All vehicles had sealed lead-acid or nickel-cadmium batteries, ac motors, and electronics of the type expected to be commercially available by 1998. We varied the energy capacity (kwh) and power rating (kw) of batteries and ultracapacitors depending on all-electric range and acceleration. Because the ultracapacitors weighed only 100 pounds, the engine-electric hybrid weighed about the same as a comparable gasoline ICE vehicle; the other simulated hybrids weighed an extra 400 pounds, the pure electric about 800 more.

The hybrid engines were sized to meet the power requirement of a vehicle at maximum speed. For the range-extended battery hybrid, max-imum speed was set at 50 mph to represent urban non-freeway driving, and for the others, 65 mph. Engine power ratings varied from 6 to 25 kw, less for the range-extended hybrid, more for the other two. The range-extended and dual-mode hybrids were given all-electric ranges of 50 to 60 miles, the engine-electric hybrid an all-electric range of close to 0. We cal-culated energy use and emissions for vehicles on federal urban and highway driving cycles.

Energy Consumption

Several important facts emerge from the estimates in Table 6-1. First, vehicles operating in the hybrid mode consume about 15 to 35 percent less gasoline than comparable ICE vehicles, assuming that future ICE com-pacts have a fuel economy of about 35 mpg on the urban cycle and 46 mpg on the highway driving cycle. The improved fuel economy of hybrids re-sults from smaller engines and, in the case of series hybrids, operation at optimal speed. These fuel economy ratings may be virtually irrelevant, however, for estimating the petroleum consumption of range-extended and dual-mode hybrids. The reason is simple: if a hybrid has an all-electric driving range of 60 miles, then its engine would rarely be needed. (According to one estimate, only about 5 percent of vehicle travel occurs after a daily distance of 60 miles has already been covered.)[13]

TABLE 6-1
Energy Consumption of Hybrid Compact Cars

Vehicle Type	Urban			Highway		
	Electricity (wh/km$_{bat}$)^a	Fuel (mpg)^b	Total (kwh/km)^c	Electricity (wh/km$_{bat}$)^a	Fuel (mpg)^b	Total (kwh/km)^c
Electric	116	—	0.53	103	—	0.47
Range-Extended						
Battery	—	46	0.58	24	72	0.48
Dual-Mode	—	46	0.58	—	53	0.51
Engine-Electric^d	—	52	0.52	—	56	0.48
Gasoline ICE	—	35	0.77	—	46	0.58

^{a} Electricity consumed by the vehicle measured at the battery (or ultracapacitor).

^{b} Miles per gallon of fuel consumed while the vehicle is operating in the hybrid mode.

^{c} Total primary energy consumed in the entire fuel cycle, taking into account energy losses at the re-finery and power plant, during distribution, and on board the vehicle. A power plant efficiency of 33 per-cent is assumed.[12]

^{d} In this analysis, the fueled engine-electric hybrid uses ultracapacitors, not batteries, for on-board electricity storage and peak power.

The second finding, therefore, is that the gasoline consumption of a range-extended battery or dual-mode hybrid with extended all-electric range would be less than 10 percent that of a comparable gasoline-powered ICE vehicle.

Third, the engine-electric hybrid is also more energy efficient than comparable ICEs—but the gasoline savings would not be as impressive as those of other hybrids. Unlike the others, this hybrid derives all its energy from gasoline (or other chemical fuel). Amory Lovins argues that an engine-electric hybrid such as his supercar could attain fuel economies several times higher than indicated here.[14] His much more ambitious estimates result from the assumption that tires and accessories would be extraordinarily energy efficient and materials lightweight. By reducing vehicle weight to about 1,000 pounds, he not only reduces energy needs but also allows further downsizing of batteries, engine, and motor, resulting in still greater energy savings.

Fourth, the hybrids and pure electric vehicles all consume about the same amount of total energy. Efficiency gained by range-extended battery and dual-mode hybrids as a result of lighter battery packs is more than offset by the extra weight and inefficiency of the engine. The engine-electric hybrid suffers from a fundamentally inefficient energy conversion process (chemical to mechanical to electrical and then back to mechanical), but benefits from optimal control and engine use.

Emissions

Hybrid tailpipe emissions can be much lower than those from ICE vehicles, but only in the most extraordinary circumstances are they likely to match the full fuel-cycle emissions of a pure electric vehicle. Three strategies would yield tailpipe emissions significantly below those of any gasoline-powered vehicle: extend the all-electric capabilities of vehicles so that they can operate as ZEVs most of the time; when the engine is on, run it at near-constant speed and power (eliminating wildly fluctuating engine conditions); and substitute a very-low-emitting gas turbine or direct-injection methanol-diesel engine in the series hybrid.

With emission-control technology as sophisticated as that on gasoline cars, hybrid vehicles should readily meet California's stringent "ultra-low-emission" rating when operating in the hybrid mode (see Table 6-2). Average tailpipe emissions from range-extended and dual-mode hybrids would be much lower than from gasoline-powered vehicles because these hybrids would operate in an all-electric mode most of the time.

Because emissions from an engine are generally much higher per vehicle mile than those from a power plant, it is important to determine

TABLE 6-2

Emissions from a Hybrid Compact Car in an Urban Setting
Using Advanced Emission-Control Technology (grams per mile)

	Power Plant Emissions in All-Electric Mode[a]	Tailpipe Emissions in Hybrid Mode[b]	Total Emissions, Range-Extended and Dual-Mode Hybrids	Total Emissions, Engine-Electric	ULEV California Rating[c]
Hydrocarbons	0.005	0.05	0.007	0.05	0.04
Carbon monoxide	0.04	0.45	0.06	0.45	1.7
Nitrogen oxides	0.15[d]	0.46	0.17	0.46	0.2

[a] Adapted from Q. Wang, M. A. DeLuchi, and D. Sperling, "Emission Impacts of Electric Vehicles," *Journal of the Air and Waste Management Association* 40: 1275–84 (1990); and H. Dowlatabadi, A. J. Krupnick, and A. Russell, "Electric Vehicles and the Environment: Consequences for Emissions and Air Quality in Los Angeles and U.S. Regions" (Resources for the Future Discussion Paper QE91-01, Washington, D.C., October 1990). Emissions include those from power plants outside the air basin and correspond to an average power-plant mix in the United States, fairly well controlled power plants, and moderately efficient vehicles.

[b] Adapted from A. F. Burke, "Development of Test Procedures for Hybrid-Electric Vehicles" (EE&G report DOE/ID-10385, 1992). Tailpipe emissions in the hybrid mode are averaged over all miles in urban driving. About a quarter of total vehicle mileage is defined as nonurban, much of which would need to be traveled with the engine on if a hybrid vehicle were used. ZEV range is 60 miles; range in the hybrid mode is 300 miles.

[c] These are standards for 50,000 miles. The standards at 100,000 miles are somewhat higher.

[d] Nitrogen oxide emission estimates for electric vehicles can vary considerably from one region to another.

how frequently and how long the engine is used, how often and to what extent the vehicle's battery range is exceeded, and when the battery is recharged.[15] Elsewhere, A.F. Burke developed a method for estimating the average emissions of hybrid vehicles; he drew on the same vehicle simulation model used above to estimate electricity and fuel consumption.[16]

Burke estimated vehicle emissions with data from steady-state tests of gasoline engines. Emissions from hybrid engines should be easier to predict than emissions from conventional vehicles because hybrid engine speed varies slowly, and because abundant electrical energy is available on board to preheat catalysts and reduce high cold-start emissions.

The following conclusions can be drawn from Table 6-2. First, hybrid vehicles would emit much less in the all-electric than the hybrid mode. Second, a hybrid vehicle operating in the hybrid mode, with both engine and motor on, may generate hydrocarbon emissions similar to those of an ultra-low-emission vehicle, but possibly with higher nitrogen oxide emissions. Third, average emissions (including from the power plant) for hybrid cars appear to be much lower than for gasoline cars. Fourth, only in exceptional circumstances would a hybrid vehicle match the full fuel-cycle emissions of a pure electric vehicle. Only if the hybrid were to operate virtually all the time on stored electricity (but then why bother with a hybrid?), or if pollution rates were high at local power plants (thus making the on-board ICE appear relatively clean), would the hybrid's emissions approach those of a pure ZEV. It is especially difficult to imagine an ICE on the hybrid vehicle matching the virtually nonexistent hydrocarbon emissions associated with electricity generation. In California, where power plants are very clean, do not burn coal, and are often located outside the air basin, hybrid vehicles have essentially no chance of equaling the emissions of pure electric vehicles, and therefore no chance of gaining ZEV credits as currently defined.

Air quality regulators question whether emissions from hybrids would be as low in practice as in theory. The hybrid engine might run quite a bit more in the real world than the simulations imply. Drivers might depend more on their engines because they don't recharge the battery frequently enough or because they retained the battery pack beyond its prime. Or drivers might override the computer controls, using the engine to gain more power and thereby increasing emissions severalfold. Moreover, engine emissions tend to be much higher under actual conditions than in simulations and laboratory tests—a result of poor engine and emission-control maintenance, frequent on-off cycling of the engine, and longer and harder accelerations than anticipated.

It is difficult to know to what extent this skepticism is justified. On the one hand, no evidence is available to determine whether the engine would be used more than expected, for the simple reason that extended market and consumer tests have never been conducted with state-of-the-art hybrid vehicles. Meanwhile, evidence *is* available showing higher in-use emissions. It is now well known that in-use hydrocarbon and carbon-monoxide emissions from gasoline engines may be up to four times higher than tested emissions.[17] This evidence, plus the unique hybrid phenomena of frequent engine start-ups and cold starts, does cast doubt on the low emission estimates. On the other hand, it is much easier to reduce emissions from engines operating at constant speed (the case with hybrid engines) than from engines operating at rapidly varying speeds (the case with ICEs).

Costs and Marketing

Like pure battery-powered electric vehicles, hybrids will initially cost more than conventional ICE vehicles, primarily because gasoline vehicles and their components are mass produced and, so far, hybrids are not. But if production of hybrids is scaled up and improvements are made, costs will come down to a competitive level.

As we have seen, analysts are coming to believe that an electric vehicle, minus the batteries, will eventually cost somewhat less than a comparable gasoline-powered vehicle. In a hybrid, some or all of this advantage would be lost in paying for the engine and extra electronic controls. But the hybrid engine, smaller than its counterpart in a conventional vehicle of the same performance, would not be exorbitantly expensive. The small engines presently produced in large quantity for lawnmowers, tractors, snowmobiles, and boats could be engineered for hybrid applications and produced in quantity at low cost. Even the cost of the electric-hybrid control would be small, thanks to advances in microprocessor controllers and associated sensors.

Adding in the battery cost would probably make most hybrids more expensive than gasoline vehicles. It is plausible, though, that the life-cycle cost of hybrid vehicles, taking into account longer battery life and low energy consumption, will eventually approach that of gasoline-powered vehicles. And if energy and environmental externalities of vehicles and fuels are factored in, hybrids should actually prove less expensive than gasoline vehicles.

Markets and Strategies for a Middling Option

Hybrid vehicles face the same start-up problems as other electric-propulsion technologies. They utilize new technologies—motors, power electronics, traction batteries, peak-power devices—that are not currently produced in high volume for the automobile market. Like pure electrics, hybrids are unfamiliar to vehicle producers, dealers, maintenance shops, and consumers. Only with incentives or mandates would automakers produce the vehicles and consumers buy them, and incentives are not likely to be available for hybrids in the near term, mostly owing to the peculiarities of the regulatory process. Hybrids are a middling option for both regulators and automakers. Why is this?

First, hybrid vehicles, although potentially very low polluting, do not qualify as ZEVs. This is a critical point. The ZEV mandate dwarfs all other incentives in motivating automakers to produce electric vehicles. Second, automakers continue to believe that they can meet even the most stringent tailpipe emission standards with gasoline, so that not even lower emissions are an incentive to produce hybrids. If efforts to reduce emissions from gasoline-powered cars to ultra-low levels eventually falter, automakers will rethink their commitment to gasoline, but that point is unlikely to be reached until the late 1990s when the hybrid option may appear less attractive than emerging battery and fuel cell technology. Third, the stagnation of CAFE standards has reduced pressure on automakers to improve fuel economy. Automakers can easily meet the standards without resorting to new energy technologies such as those represented in hybrid vehicles. The net effect of these regulatory circumstances, even with substantial government funding of hybrid research by automakers, is to postpone until the end of the decade virtually any incentive to produce such vehicles.

Hybrids run the risk of remaining a middling option in the future. They may be kept out of the small car market by improved batteries and ultracapacitators and out of the market for larger vehicles by fuel cells. Simpler, cleaner, more energy efficient, and potentially cheaper fuel cells may be the decisive blow against ICE-battery hybrids.

Only if fuel-cell technology falters, hybrid development picks up speed, and battery development is sluggish will the passenger car market open up to hybrids. Even then, it would be necessary to overhaul the regulatory system—to create flexible, incentive-based regulations that fully rewarded hybrids for their energy and environmental advantages.

The auto industry is at loggerheads with California regulators over hybrid vehicles. Major automakers, fearful that limited-range electric vehicles will not sell, would like to fulfill at least part of the ZEV mandate with hybrids. CARB has said that it would be willing to create a separate (additional) emission category for hybrids. It has also expressed willingness to count hybrids as ZEVs if the full fuel-cycle emissions of hybrids were less than those of battery-powered electric vehicles. The automakers find these responses unacceptable. They don't want additional new requirements, and they realize a hybrid could not beat the full fuel-cycle emissions of a battery-powered electric vehicle in California.

CARB's position is probably wise for now, at least with respect to California. Dramatic reductions in emissions are needed there to make the air healthy to breathe. A strong ZEV mandate that pushes companies over start-up barriers and accelerates the development and commercialization of inherently cleaner technology seems appropriate. In places where emission reduction is less urgent, a more gentle regulatory approach may be appropriate. In any case, flexible, incentive-based rules that take into account other energy and environmental goals should eventually replace the ZEV mandate. The sooner that is done, the better the prospects for hybrids.

Perhaps a two-pronged approach to electric drive is advisable: let California focus on pure ZEVs, while the rest of the country focuses on hybrids. Indeed, the large hybrid projects being funded by the U.S. Department of Energy and the Partnership for a New Generation Vehicle are premised on reducing energy use, not pollution. (Hybrids may also be of interest in countries where few households own more than one car.)

Whether or not hybrids succeed, expanded investment in hybrid research and development will not be fruitless. Even if fuel cells emerge as the dominant automotive technology, much of hybrid-electric technology will apply directly to fuel cell–electric vehicles. Advances in driveline technology would naturally evolve from all-electric to ICE-electric to fuel cell–electric vehicles. Indeed, the clean, highly efficient vehicles of the future are likely to be hybrids—if not ICE-electric hybrids, then fuel cell–electric hybrids.

CHAPTER 7

Accelerating Regulatory Reform

*Although we have this enormous set of environ-
mental laws, they are all anecdotal and separate . . .
and none of them states an overall long-term goal.*

Jonathan Losh, President, World Resources Institute

The regulation of vehicle emissions, energy use, and safety calls for fundamental change. The current system is too inflexible and inefficient to
handle the growing diversity of technological options. It inhibits shifts toward very small vehicles, new propulsion systems, and new fuels. Part of
the problem is the adversarial relationship that has evolved between regulators and fuel and vehicle suppliers. A far more flexible and incentive-
based approach is needed to navigate the complex transition to a more benign transportation system.

An example of the regulatory nightmare confronting those who
would introduce new vehicles and fuels is the arcane system of building
codes.[1] No single set of model codes prevails across the United States.
Each locality adopts codes at its own pace (a process often taking years)
and enforces them seemingly at random. Conservative interpretation of
building codes could disrupt installation of recharging units in homes
and elsewhere and stymie the acceptance of electric vehicles. The result
would be higher prices for electric vehicle charging and long delays in installing charging equipment. The following problems might arise from
building code requirements applied too stringently or inappropriately:

- Because of toxic materials in some batteries, electric-vehicle
  chargers could be classified as hazardous, resulting in strict limitations on the height and area of the building and a ban on chargers in
  residential garages.

- Charging could be banned, or expensive "watertight" charging
  equipment required, if located in a floodplain.

- Charging, if classified as vehicular fueling, could be prohibited in
  commercial garages because of rules regarding the transfer of
  liquids.

- Special fire-sprinkler systems could be required in parking structures.

- Charging could be banned in enclosed garages if it were deemed necessary to ventilate battery gases.

- Expensive drainage systems could be required for toxic and corrosive battery materials.

- Charging could be limited at commercial garages if electric cars were classified as spark-producing vehicles.

The point of this list is not to suggest that electric vehicles are a major safety hazard. They are not, and in fact should prove at least as safe as gasoline vehicles. The point is that, to protect us from the failings of the marketplace, government has created highly specific and rigid rules and regulations which may have been appropriate in the past, but which are proving inappropriate for emerging vehicle fuel and battery technology.

The Current Regulatory Approach

The 1960s ushered in an era of government activism in transportation. Regulatory programs were adopted to improve vehicle safety—hastened by Ralph Nader's book, *Unsafe at Any Speed* (1964)—and to reduce air pollution. Fuel-use regulations followed in the 1970s.

The regulatory system that evolved was a creation of lawyers and engineers, whose professions are founded on highly specific rules of conduct and design. The result is an approach that has come to be known as command and control.[2] While regulations affecting vehicles and fuel are not strictly command and control—they contain some flexibility—most tend to be directives that restrict the behavior of vehicle and fuel suppliers.

Regulation of Vehicle Emissions

The process of regulating vehicle emissions is highly structured. Each year automakers receive stacks of documents from the EPA for vehicles sold in forty-nine states, from CARB for those sold in California, as well as from the European Economic Community and Japan for vehicles sold abroad. The documents contain minutiae on the design and conduct of emission tests and advise of changes in reporting requirements, test procedures, and control technology. Each and every car, bus, and truck must meet the uniform standards laid out.

Emissions are measured and tested on a grams per mile basis (except for evaporative hydrocarbons from the fuel tank and elsewhere, which are tested separately). Compliance is verified by running sample vehicles through a standard driving cycle. Initially, emission standards were established for three pollutants: carbon monoxide, hydrocarbons, and nitrogen oxide. Standards have been tightened intermittently based on subjective judgments of the severity of ambient pollution and of what standards manufacturers are capable of meeting at reasonable cost. In a radical departure from uniform standards, California instituted the idea of emission averaging, trading, and banking in 1994. (As indicated earlier, California is the only state allowed to create its own regulatory program for emissions. Other states may adopt either the California or the federal program, and until 1993 all of them opted for the federal program.)

The strength of the uniform-standard approach, which should not be minimized, is simplicity and apparent ease of enforcement. But this approach is fundamentally flawed, and becoming more so as new fuels and vehicles enter the marketplace. Consider, for example, vehicles powered by fuel cells, batteries, biomass fuel, and natural gas. Today's system for regulating emissions addresses only those emissions from the vehicle. "Upstream" emissions—from power plants, oil refineries, fuel stations, and so forth—are ignored, even though for these new fuels and vehicles they tend to be larger than "downstream" tailpipe emissions. Even for gasoline vehicles, refinery and fuel station emissions may soon be as high as vehicle emissions. The ZEV mandate in California solves the problem of upstream emissions by simply assuming that power plant emissions are zero—an approach arguably acceptable for kickstarting the electric vehicle industry, but clearly not acceptable for the long term.

Unless power plant emissions (and the upstream effects of biomass and other nonpetroleum fuels) are considered, the truly zero-emitting fuel cell vehicles of the future, those operating on solar hydrogen, would not be distinguished from battery-powered electric vehicles drawing electricity from, for instance, dirty coal-fired power plants near the Grand Canyon. A new regulatory approach that addresses emissions over the full fuel cycle is urgently needed.

A second failure of today's rigid and fragmented standards is that they exclude engine technologies and fuels that could provide substantial pollution and energy benefits, but that are unable to meet the emission standard for a particular pollutant. Even innovative internal combustion techniques such as "lean burn" and two-stroke engines, which appear to be in the public interest, are hindered by uniform emission standards.

In lean burn as opposed to conventional combustion, less fuel is mixed with air. This results in more complete combustion and improved fuel efficiency—about 5 to 10 percent more efficient in the case of Honda's 1992 Civic VX.[3] There are also fewer carbon dioxide, hydrocarbon, and carbon monoxide emissions. Because of higher combustion temperature, nitrogen oxide emissions tend to increase. The current regulatory system has no mechanism for trading the advantage of less fuel consumption and overall emission reduction with the disadvantage of somewhat higher nitrogen oxide emissions. This makes it difficult, if not impossible, to introduce lean burn engines, especially as nitrogen oxide emission standards are tightened over time. It is the reason Honda did not sell lean burn cars in California in the early 1990s, even though other emissions and fuel use were far lower for these than for most other cars.

The story is similar for two-stroke engines.[4] They provide greater fuel efficiency than conventional four-stroke gasoline engines, but with somewhat higher nitrogen oxide emissions.

Rarely is there a new strategy or technology that is uniformly positive across the board. A more flexible framework would allow trade-offs between different pollutants and different energy and environmental goals.

A third shortcoming is the economic inefficiency of uniform standards. Uniform standards are not sensitive to differences in the cost of reducing emissions from one vehicle to another. Worse, because they offer no incentive to reduce emissions below the standard, manufacturers have no financial reason to introduce cleaner-burning engines and fuels. In fact, there is a strong incentive *not* to innovate. Automakers rightly fear that any improvements they make will soon be used against them—that those specific improvements will be either explicitly required by regulators or used as a justification for further tightening of standards.

This disincentive inevitably creates an adversarial relationship between industry and regulators. Industry sees government regulators as trying to take advantage of them with rigid prescriptions, while regulators feel that they cannot trust industry to be straightforward about the cost and difficulty of lowering emissions and energy consumption and improving safety. This situation can only exacerbate mistrust and ill will between the parties.

The disincentive to innovate is not absolute, nor is the auto industry monolithic. Occasionally a particular company will aggressively pursue a much cleaner or more efficient technology with the hope of winning a market or public relations advantage. The development and unveiling of General Motors' Impact in early 1990 is an example of such occasional breakaway initiative. Indeed, that unveiling, which the company later came to regret, led directly to California's adoption of the ZEV mandate in late 1990. Another initiative is Ford Motor Company's cancellation of a proposed research program to redesign the Taurus with an all-aluminum body. Reportedly, the company feared that regulators would use the findings to justify more stringent fuel economy standards for all vehicles.

The shortcomings of the current approach toward regulating emissions can not be ignored, if only because of its reach and influence. The goal of reducing air pollutant emissions from motor vehicles has dominated other policy initiatives in the transportation sector, including not only the introduction of new fuels and vehicles but also decisions regarding urban land use, transportation infrastructure, telecommuting, transit, carpooling, and even employee work hours. This situation is the result of twenty-five years of strong popular support for clean air and for the creation of specific and enforceable rules to reach that goal. In essence, the air-quality tail wags the transportation dog—sometimes to good effect, sometimes not.

The entire regulatory system needs to be redesigned. An effective system would accommodate differences among regions, fuels, and engines, provide incentives to industry to innovate and to consumers to buy cleaner vehicles, and allow trade-offs among social goals. The present system fails on all counts. Because of the political support for, and rigidity of, motor vehicle regulation, it sometimes even perversely results in regulations that deter the introduction of environmentally benign vehicles and fuels. While clean air regulation has resulted in many positive changes, including rules that jumpstart electric propulsion technology, a more rational and balanced framework is needed for the long haul.

Fuel Economy Regulation

U.S. standards for fuel consumption are somewhat more flexible than those for emissions, but they too have their shortcomings. Instead of meeting a uniform standard for all vehicles, the manufacturer complies with a standard reflecting average fuel consumption across all vehicles it sells in a particular year. The CAFE standards, created in 1975 and made effective beginning in 1978, allow manufacturers to build a mix of vehicles, some larger with powerful engines, others smaller and more fuel conserving. The CAFE standard for cars, set regularly by Congress, reached 27.5 mpg in 1985, was pushed back to 26.0 for a short while, and remains deadlocked at 27.5 as this is written. The CAFE standard for light trucks, set by the U.S. Department of Transportation, was creeping up at about 0.1 mpg per year through the early 1990s, and neared 21 mpg for the 1996 model year.

Though premised on averages, CAFE standards are still not very flexible. Complex rules devised to distinguish imported vehicles from those produced in the United States and to separate light trucks from cars, has led to many abuses of the spirit of the program. (The classification rules were created mostly to protect the domestic Big Three from Japanese competition.) For example, at one point Ford reportedly transferred the manufacture of some minor components of its gas-guzzling Crown Victoria to Mexico so that the car would count as an import, thus allowing the company to produce more large cars and still be within the 27.5 mpg limit for domestic cars.

Moreover, CAFE averages hurt manufacturers that specialize in gas-guzzlers such as sports cars, luxury cars, and recreational vehicles. That in itself is not necessarily undesirable. The problem is that it offers no *rewards* to manufacturers, small or large, that specialize in fuel-efficient cars.

The principal weakness is that the CAFE approach is out of synch with market forces. Automakers vigorously oppose increases in the average standard because they insist, for the most part justifiably, that consumers are not willing to pay extra for lower fuel consumption. For their part, consumers are acting rationally. With gasoline costs lower than they have been since World War II (taking into account inflation), an additional 5 mpg generates less than a hundred dollars in fuel savings per year.

What worked in the past will not necessarily work in the future. Fuel economy regulation had some effect during the late 1970s and early 1980s when fuel prices were high. High fuel prices encouraged manufacturers to invest in more efficient vehicles and consumers to buy them, and the schedule of stiffening fuel economy standards helped instill a sense of security in the face of an uncertain future. High fuel prices and stiffening standards reinforced each other.[5] That dynamic has unraveled with sagging fuel prices and the increasing cost of ICE fuel economy improvements.

The Light Truck Story

The rigidity of today's regulatory system and the difficulty of altering it to accommodate new technologies is illustrated by government's gentle treatment of light trucks. In the mid-1970s, light trucks such as minivans, four-wheel-drive recreational vehicles, and pickup trucks accounted for only about 10 to 15 percent of light-duty-vehicle sales. By 1993, light trucks had seized 38 percent of the market.[6] Their success in grabbing sales from cars derived in part from the relatively low cost of complying with the lax emission, energy, and safety standards for light trucks. Manufacturers could provide power and performance for a good price.

The argument for not imposing more stringent standards on light trucks is that light trucks serve multiple uses, carrying goods and pulling trailers as well as transporting passengers. This is a specious justification, since the vast majority of light trucks are used strictly for passenger travel. Lacking political will and analytical expertise, regulators deferred to the automotive industry on this score, establishing much weaker standards for light trucks than for cars. To this day, light trucks have less stringent CAFE, emissions, and safety standards, even when they are roughly the same size as cars and serve the same function.

Under the current regulatory system, a distinct set of procedures and rules is established for each new type of vehicle, propulsion system, and fuel. Each change in rules causes a clash between regulators and automakers. Is this situation desirable or even tenable for the future? Would regulators be willing (or equipped) to do battle against the auto industry over every new vehicle and fuel that makes its debut in an era of rapidly diversifying technologies and energy sources? Probably not.

Safety Standards

Safety standards, more rigid than emission and fuel economy standards, are a combination of technological and performance standards. Technological standards mandate specific types of glass, physical restraints, and other devices in vehicles. Performance standards are somewhat less restrictive: they establish goals with respect to such issues as survivability in prescribed test crashes using dummies. Technological and performance standards have had little effect on emissions and, for the most part, fuel economy. They have also had little effect on the design of conventional-sized electric vehicles. However, they have had a major influence on the design and use of small vehicles.

Safety regulators singlemindedly pursue a narrow interpretation of their safety goals. They focus on vehicle design features needed to ensure survivability in collisions between vehicles at high speed. They make extensive use of specific technological standards and unquestioningly accept conventional-sized cars as the norm. This approach is reasonable so long as vehicles and road systems are fairly homogenous. But it shuts out new vehicle-road designs that might accommodate smaller, more environmentally benign, and less expensive cars, even those such as neighborhood electric vehicles that might operate in a safer setting.

Regulators adhering to the status quo have a legitimate retort: How could they justify allowing the sale of small vehicles, which might be less safe than larger vehicles, without any assurance that the small vehicles would be used in a sheltered environment? But it is precisely that view that freezes efforts to introduce smaller vehicles that are not only environmentally and economically superior, but also provide greater mobility to many. As indicated in Chapter 4, these small vehicles may represent the first step toward safer local travel.

This narrow approach to safety forced the Station Car Association, a group of electric utilities and transit operators organized in 1993, to abandon plans to procure very small vehicles for commuters traveling between home and transit stations, and between transit stations and workplace. The association backed down out of fear of liability, because the vehicle would not meet strict Federal Motor Vehicle Safety Standards.

Safety standards, although highly successful in improving overall passenger safety, should be adapted to a larger range of vehicles. A more flexible system is needed that covers smaller cars while assuring that their use would lead to an overall reduction in injuries and fatalities. The challenge is to create a safer transportation *system*, not just safer vehicles. Safety regulation needs to improve safety by treating it in a broader context.

Public Utility Commissions

Motor vehicles and transportation fuels are now becoming subject to additional layers of regulation: state and local public utility commissions. These governmental bodies regulate electricity supply companies that have been granted monopolies for prescribed geographical areas.[7] The commissions' primary mandate is to protect ratepayers against electricity (and other) monopolies by carefully regulating their activities, especially the prices they charge.

The problem is contradictory government goals. Public utilities commissions place the financial interests of electricity (and natural gas) ratepayers over broader energy and environmental goals. They almost totally ignore air quality, energy, and other social goals in their deliberations, to the detriment of the public interest. Because electricity has not been used as a propulsion source until now, electricity regulators have no experience with passenger vehicles, and this only exacerbates the problem.

The California Public Utility Commission has been one of the most progressive in acknowledging the role of energy and environmental goals, but even it insists that "the cost of LEV [low-emission-vehicle] programs can be borne by ratepayers only if and when those programs are in the best long-term interest of ratepayers."[8] The commission believes that electric utilities should adhere to their traditional responsibility of providing safe, reliable, efficient service, and not take it upon themselves to encourage substantial market penetration of electric vehicles; that activity, it maintains, is the province of others.

Other public utility commissions take an even more restrictive position. A research report prepared for the National Association of Regulatory Utility Commissions argues, with respect to natural gas vehicles (but with a logic that applies equally to electric vehicles), that "ratepayers should provide funding only for the purchase of natural gas vehicles used by the LDC [local distribution company], the construction of public refueling stations for demonstration purposes in the initial few years when the NGV infrastructure is not fully developed, and when the direct benefits to ratepayers for increased gas sales due to more NGVs in use can be clearly demonstrated."[9]

The net effect of narrow commission attitudes and rulings is to discourage electric utilities from investing in electric vehicle research and development, recharging infrastructure, and demonstration vehicles. If commissions could be convinced that electric supply systems would be more efficient with electric vehicles—through off-peak charging, for example—they might be more supportive of the industry. But the absence of any regulatory framework for reconciling environmental and economic goals means that their decisions will be erratic, discouraging utilities from aggressively pursuing initiatives in this area. The net result is that electric utilities, potentially the strongest constituents for electric vehicles, are shackled.

California's Bold Initiative

On September 28, 1990, California acknowledged the shortcomings of the uniform-standards approach to vehicle emissions regulations. On that day CARB adopted what became known as the low-emission vehicle (LEV) program. This initiative contained not only the ZEV mandate discussed earlier, but also two other radical elements: dramatically reduced vehicle-emission standards, and rules allowing vehicle suppliers to average and "bank" emissions (similar to CAFE), as well as trade them.[10]

This last provision—the averaging, banking, and trading of emissions—took effect in 1994. As we have seen, each vehicle manufacturer averages emissions across all vehicles it sells in a given year. It may bank or sell emission credits if it beats the standard (see Table 7-1), or buy credits if it does not. Various limitations were imposed to simplify monitoring and enforcement. For example, averaging, trading, and banking can only be conducted for a single pollutant, with the other pollutants

TABLE 7-1
California Average Automobile
Emission Standards for
Hydrocarbons (grams/mile)

Year	Emission Standard
1994	0.250
1995	0.231
1996	0.225
1997	0.202
1998	0.157
1999	0.113
2000	0.073
2001	0.070
2002	0.068
2003	0.062

Notes: Hydrocarbon emissions are measured as reactive organic gases. These standards apply to all light-duty vehicles under 3,750 pounds, measured after 50,000 miles. Somewhat higher standards were set for 100,000 miles. The ultra-low-emission vehicle (ULEV) category is for vehicles with less than 0.04 grams per mile of nonmethane organic gases (as well as 0.2 grams per mile of nitrogen oxides and 1.7 grams per mile of carbon monoxide).

rigidly linked to that one pollutant. This emission trading provision represents the first major effort to create a "market" for vehicle emissions.[11]

The LEV program is a sharp departure from uniform standards. It offers manufacturers a direct incentive to beat the standard through innovation. Manufacturers themselves are allowed to choose the most cost-effective strategy for reducing emissions. For instance, they might focus their efforts on cleaning up gasoline engines or introducing cleaner fuels, or both. Such flexibility opens up the possibility of substantial savings. If an emission averaging and trading program had been in place in California in 1990, according to one study, the cost per vehicle for attaining the same emission reduction would have been 13 to 30 percent less.[12] (If the study had also considered banking, the savings would have been even greater.) With pollution-control costs running at about $750 per vehicle in 1990, the savings would have amounted to between $150 and $350 million *per year* for California.[13]

How did this revolutionary initiative come about? CARB found that tradable credits blunted automaker opposition to stringent new emission standards. A year-long series of workshops and public hearings revealed that the flexibility inherent in emission credits was attractive to an industry that until then had been forced to accept uniform emission standards.

Flexibility has its drawbacks, though, as CARB discovered in the years after 1990. CARB had expected that the stringent new standards would direct automakers' attention to methanol and natural gas fuels, which are inherently cleaner burning than gasoline and provide other benefits as well. That has not been the case. Instead, automakers vigorously renewed their efforts to clean up gasoline engines. One reason is lingering uncertainty over the future of alternative fuels. No automaker wanted to be stuck with a whole new set of Edsels. They preferred the less risky strategy of adhering to the status quo, even if the cost of doing so was greater than switching some vehicles to alternative fuels. Another reason is the absence of incentives and mandates for fuel suppliers to sell natural gas and methanol. This indicated to automakers that the government and the fuel-supply industry were not serious about marketing alternative fuels. Furthermore, because CARB ignored energy security and greenhouse benefits (it has no authority in these areas), the incentives for switching to alternative fuels were smaller than they should have been. Perhaps a fuller incorporation of external costs into the price of alternative fuels would have convinced automakers to invest in new fuels.

Two lessons emerge from this experience. First, the extra cost of gasoline must be very large before automakers and oil companies will risk switching to other fuels. Because CARB's only mission is air quality, it cannot justify policies and rules that encourage fuel switching on grounds other than air quality. And yet by ignoring other social goals, CARB is not acting in society's best interests.

Second, flexible rules may not provide enough incentive for companies to overcome start-up barriers. Vehicle manufacturers chose to renew their investment in gasoline even though it may not have been the most cost-effective strategy. With the industry at a crossroads, with start-up barriers large and the need for change urgent, it appears that specific government directives are necessary. Therein lies the justification for the ZEV mandate and large start-up subsidies—to surmount the many barriers to electric propulsion.

CARB's 1990 LEV initiative was extraordinary in its vision and scope. It was a huge first step away from command-and-control rules toward an incentive-based approach. It included the technology-forcing ZEV mandate but also gave industry leeway in deciding the means to the end, which is cleaner vehicles. CARB is to be commended for its creativity and perseverance, especially in view of the shortsightedness that prevails at the national level and elsewhere in the world. But it is, after all, just a first step. CARB did not establish a mechanism allowing for the development of region-specific strategies, or incorporate social goals other than pollution, or address upstream pollution and costs, or motivate energy suppliers to sell clean nonpetroleum fuels.

Toward More Flexible Regulation

Transportation and energy choices facing the United States and others are overwhelmingly complex. There is a bewildering array of possible technologies whose costs are anything but certain. No single individual or company is equipped to propose a detailed blueprint for the future. What is needed is a sophisticated governmental framework for guiding the changes that originate in the private sector.

In the past, decisions in the transportation field were simpler—or seemed to be—and cost implications were modest. It was not unreasonable to expect government administrators to have the foresight to oversee the choice of fuels and technologies. That no longer is the case. Today, not only are choices far harder to make but the implications of those choices are more far-reaching. Given uncertainty about the future, a more efficient and resilient approach than command and control would be to offer financial incentives to industry and consumers that would push the transportation sector toward less pollution and perhaps greater reliance on domestic energy sources. This strategy would change the behavior of individuals and industry by appealing to the pocketbook rather than limiting choices. It would not depend on the empty hope of government omniscience.

There are two possible incentive-based approaches: one, manipulate key variables of the existing market, particularly prices and information; and two, "create" market forces through incentives. The emphasis of both approaches is on decentralized decision making, driven by self-interest but guided by a body that regulates incentives.

Fee-Bates and Taxes

Fuel taxes are a prominent example of the first approach (i.e., to improve existing markets). They are a simple means of incorporating externalities such as air pollution and climate change into fuel prices. In theory, consumers soften the blow of the tax surcharge by purchasing more fuel-efficient vehicles and driving less. In practice, fuel taxes are unpopular and, because fuel is such a small part of the total cost of operating a vehicle in the United States (about 15 percent), taxes need to be very large to make much difference.

An example of a more promising price-altering method for influencing change is the so-called fee-bate, a program of fees and rebates for buyers of new vehicles.[14] Buyers would receive a rebate if the car they purchased had better fuel economy and lower emissions than average, or would pay a fee if the vehicle emitted more pollution or used more fuel than average. The size of the fee and rebate would depend on how much better or worse the vehicle was than average. Individuals and organizations would have an incentive to purchase cleaner burning and more fuel-efficient vehicles; vehicle manufacturers the incentive to develop and sell such vehicles. Here the principal challenge for regulators would be determining the appropriate magnitude of fees and rebates to elicit the desired behavior.

There are many variations of this sort of proposal—for example, setting fees for the amount of driving done. These hold a lot of promise, especially when packaged with other technology-based initiatives of the sort discussed in this book. But legislators and regulators are wary of policy instruments to which the unpopular "tax" label can be applied. In general, taxes and fees are most palatable when they are revenue-neutral—for example, when fees balance rebates—and when revenue is deposited in trust funds set aside exclusively for clean air and other popular goals.

Marketable Credits for Vehicle Suppliers

The second incentive-based approach—creating new market forces through incentives—includes pollution licenses and permits, and marketable credits. Licenses and permits allowing companies and other entities a specified amount of pollution are less attractive in the transportation arena than marketable credits because they tend to discourage newcomers from entering the market, they are difficult to adjust for shifting economic, technological, and political conditions, and they are difficult to assign equitably.

Marketable credits, however, show great promise, with respect not only to vehicles, as has been shown in California, but also to fuels. Marketable credits are created by setting targets. Suppliers that do better than the standard can "bank" and trade their excess credits, thus creating a market with credits as the currency.

The beauty of a marketable credit scheme is its flexibility. Manufacturers that prefer to focus on less environmentally benign fuels and vehicles—gasoline, diesel fuel, large engines, jeeps, and so forth—may continue to do so; they simply buy credits from those other manufacturers that are bettering the standard.

Flexibility translates directly to less cost by providing industry—driven above all by cost—the freedom to reduce emissions in the most cost-effective manner possible. If, as indicated earlier, California would have saved several hundred million dollars in 1990 with an emission trading scheme, the country as a whole would have saved $1 to $3 billion per year. As emission standards are tightened during the 1990s, the cost of meeting emission goals will rise, along with the savings realized by marketable credits.

The benefits of marketable credits would expand further if the automaker-regulator relationship were made less adversarial by the new incentive-based system, and if other social costs—including fuel efficiency, air toxics, greenhouse gases, and energy security—were brought under the marketable-credit umbrella. Incorporating additional costs would also have the advantage of greatly increasing the dollar value of credits for cleaner and more efficient vehicles.

Marketable Credits for Fuel Suppliers

Like most other regulatory initiatives to reduce pollution and energy use in the transportation sector, California's LEV program, including the ZEV mandate, placed the burden squarely on motor vehicle manufacturers. With the notable exception of lead removal, regulators have until now mostly ignored fuel supply. As fuel choices enter the air quality and climate-change policy calculus, this situation is no longer tolerable. Extending marketable credits to fuels may not be what the oil industry wants, but it could lead to a less costly and more effective transition to cleaner fuels and vehicles. This concept has already been implemented with power plants; the 1990 Clean Air Act Amendments enacted a program for trading sulfur oxide emissions that is now active and proving less costly than previous command-and-control techniques.

In a speech in July 1989, President George Bush finally shifted the environmental debate to transportation fuels. His proposal to substitute methanol for gasoline to meet clean air goals alarmed the oil industry. Atlantic Richfield (Arco), a regional oil company headquartered in Los Angeles, along with its industry brethren, launched a counterattack and soon unveiled their preferred substitute: cleaner-burning reformulated gasoline. Air pollution regulators in Washington, D.C., and Sacramento quickly embraced the proposal, adopting stringent new gasoline standards that are essentially prescriptive in spirit.

The conversion of this recent set of gasoline rules into a broader marketable credits program would have three considerable advantages not otherwise available. It would make room for region-specific strategies, incorporate upstream impacts, and lower emissions from existing gasoline-vehicle fleets.

Fuel regulation lends itself to region-specific strategies because virtually all the fuel purchased within a region is consumed within that same region. But it would take some effort to administer a fuel regulation program. It would be more difficult than a comparable program for automotive emissions, principally because there are many more fuel suppliers than vehicle suppliers and because several major fuel supply industries—natural gas, oil, electricity, and perhaps even agricultural suppliers of biomass fuels—would have to be accommodated. Furthermore, regulators have little experience with transportation fuels (though they are familiar with the refineries and power plants that produce the fuels).

Compounding all this is the need for a method to account for the differing quantities and types of pollutants emitted at different stages of the fuel cycle, differences that vary from region to region. The challenge can be met by creating different standards in different regions for attributes of each fuel. Regulation of fuel attributes, such as greenhouse gas or nitrogen oxide emissions, readily lends itself to the development of region-specific strategies. The standards can be adjusted to reflect the pollution "fingerprint" of a given area, and the average standard required of each fuel supplier can be raised or lowered depending on the severity of the problem in that area.

In Los Angeles, for instance, the average emission standard fuel suppliers would meet for nitrogen oxides might be 0.1 grams per mile by the year 2005, while in cleaner San Francisco it might be a less stringent 0.3 grams. In fuel-based regulation, each supplier would determine the most cost-effective manner for meeting specified average ratings. If it were expensive for an oil refiner to reduce emissions by reformulating gasoline, perhaps owing to the design of its refineries, or if the average standard were lower than what is achievable with reformulated gasoline, then credits could be purchased from a natural gas or electricity company that met the required standard at less cost. Or the oil refiner might choose to sell natural gas or even electricity at its own stations.

An important step would be taken by designing the fuel standard to reflect other social goals, such as reduced emissions of greenhouse and toxic gases and greater energy security. Ideally, fuel suppliers would meet these goals in the same way that they meet emission standards. For instance, the standard for greenhouse gases could be set at 1.0, with ratings assigned to each fuel depending on its effect on climate change. Methanol and gasoline might be rated at, say, 1.0, while lower-emitting options such as domestic natural gas might be 0.9 and hydrogen made from natural gas, 0.6.

CARB considered a simplistic marketable-credit scheme for fuels in an early version of the LEV program. As initially proposed, gasoline suppliers would have had to sell a specified amount of methanol and liquefied petroleum gas, based on sales of alternatively fueled vehicles and on their share of total fuel sales. Refiners were to be allowed to satisfy the clean liquid-fuel requirement by either selling those fuels directly or by buying credits from other suppliers who had sold clean liquid fuels in excess of their requirement. There would have been banking of credits, with sharp discounts over time. Up to 10 percent of the fuel-sale requirement could have been met by CNG or electricity sales to motor vehicles. Credits were to be allowed only for electricity sales above the mandated ZEV level (2 percent in 1998, 5 percent in 2001, and 10 percent in 2003). In the end, most of this proposed clean-fuel program was dropped because of opposition by oil refineries, which claimed that they should not be held responsible for the willingness of consumers to purchase alternative energy.

Where Industry Stands

Those attempting to craft a successful regulatory program must confront the sobering realization that the industries most affected—auto, oil, natural gas, and electricity—are the largest and most powerful in the world, and that inertia supports current technology and investments. The challenge is to create an evenhanded regulatory format that is effective and economically efficient, and that respects the interests of those companies while pursuing the public interest. This is a daunting task.

There is some room for optimism. Interaction among the four industries, although minimal in the past, is now expanding. The U.S. ABC, a collaboration of electric utilities, automakers, and the federal government to further battery development, is one example. Another is the $30 million cooperative study of reformulated gasoline and methanol emissions by three automakers and eleven oil companies in the early 1990s.

Electric companies are entering the vehicle market with exceptional caution. They have little or no marketing expertise, have virtually no knowledge of the transportation market or the political and economic matrices within which it operates, and are prohibited from the market in some cases by regulators. With the coming deregulation of electricity supply, the utilities and their regulators are exploring new relationships. How that will play out is uncertain.

Some electricity regulators are taking the broader view. In 1993, the

California Public Utility Commission issued an opinion that explicitly endorsed broader air quality and energy goals, acknowledging that ratepayers are also citizens and residents seeking a better world. The president of the commission went further, strongly backing electric vehicles:

> Off-peak recharging of electric cars is a perfect example of efforts that should be undertaken that could provide a triple win scenario. Ratepayers would have the fixed system costs spread over a larger sales base thereby reducing per unit costs; shareholders benefit by establishing profit incentives for the sale of off-peak electricity; and society as a whole benefits by addressing one of the most vexing problems we face in California, that of air emissions from internal combustion engines.[15]

Oil companies—with huge investments in refineries, pipelines, and storage depots—have deeply entrenched interests in the transportation market. Gasoline, diesel, and jet fuel remain the heart of their business, and they have little experience with nonpetroleum fuels. With the trimming of corporate staffs in the early 1990s, their ability and willingness to consider new energy options shrank.

In general, the oil industry is more receptive to natural gas vehicles than to other alternatives, for the simple reason that they constitute the smallest risk. Relatively little new capital investment is needed, and oil companies are already heavily invested in natural gas worldwide. If they are more skeptical of methanol it is because the risk is much larger; they would need to invest $1 billion or so in each methanol plant, with no guarantee of sales. Natural gas is cheaper than gasoline, and once vehicle owners buy or convert to a natural gas vehicle, they become a captive market. Not so with the more expensive methanol, unless vehicles operate on fuel cells rather than flexible-fuel ICEs. The least attractive option for oil companies is electric vehicles; once they are perceived as a significant threat, the response of oil companies will undoubtedly become more vitriolic than their response to methanol in 1989. Indeed, in proceedings before the California Public Utility Commission in 1994, a lobbying organization representing oil companies argued strongly against allowing utilities to provide battery subsidies to electric vehicle buyers.

Oil companies headquartered in California, bombarded daily by environmental regulators seeking to ban diesel fuel, clean the skies, and electrify cars, tend to be most responsive to changes in the transportation energy market. But even they haven't made significant investments in nonpetroleum fuels. What they have done is invest billions of dollars to upgrade oil refineries to produce reformulated gasoline—as much as a billion dollars per refinery.

Whether oil companies would acquiesce to more incentive-based regulation is uncertain. They would be more likely to accept fee-bates than fuel taxes and marketable credits because fee-bates affect car purchases, not fuel purchases. Because marketable credits leave considerable discretion to fuel suppliers, it is conceivable that the concept could be crafted to gain industry acceptance.

The auto industry, not surprisingly, is less antagonistic to nonpetroleum fuels than the oil industry. All else being equal, automakers prefer as little change as possible. In descending order, they prefer methanol, natural gas, and electricity. They would rather spend billions of dollars a year to gain small additional reductions in gasoline emissions, thereby meeting emission requirements in California, than invest billions in natural gas or methanol or new electric powertrains.

However, competition among automakers is more intense now than at any time since World War II. From World War II until the mid-1970s, the U.S. auto market was dominated by a few domestic companies. They behaved as an oligopoly, reluctant to take the lead in introducing major new technological or design changes. That changed with the growing success of Japanese companies. General Motors' ambitious Impact electric-vehicle program reflected the intensified competition; it was a bold attempt to recapture market share in the small-car market from the Japanese, and to embellish General Motors' flagging reputation in California and elsewhere. Such a bold and expensive deviation from the norm would probably not have been made in the 1960s.

The greater competitiveness of the auto industry bodes well for the future. It suggests more willingness to innovate and take risks in pursuing promising new technologies. This increased competition, along with the development of flexible manufacturing techniques that are less dependent on economies of scale, encourages large companies to pursue niche products, and creates opportunities for small and inexperienced outsider companies to develop new products and vehicles in conjunction with large automakers. Indeed, major innovations such as neighborhood electric vehicles and manufacturing processes centered around lightweight materials are likely to come from outsider and small companies. In today's competitive market, though, changes will be closely monitored by the large automakers and snapped up if they appear promising.

Regulation of the auto has intensified since the 1960s. The future implications of still more stringent regulations are difficult to discern. On the one hand, familiarity with regulation eases apprehension about (and knee-jerk hostility to) new types of regulation. Resentment undoubtedly lingers, but it does not seem to dominate the industry as a whole. Indeed, the positive response to emission trading in California suggests that, for the most part, automakers will be supportive of continuing efforts to create more flexible and incentive-based formats.

The type of regulation automakers adamantly oppose is that which is at cross-purposes with price signals. They oppose CAFE standards because, with current fuel prices, energy efficiency reaps no profit; that is, consumers are not willing to pay for the higher cost of building more energy-efficient vehicles. Under heavy pressure to produce cleaner, more efficient cars, automakers much prefer initiatives that provide incentives to consumers to buy "green" cars, and that reward companies for selling them.

- - - - - - - - - - - - -

The commercialization of more environmentally friendly fuels and vehicles will be delayed and investment decisions distorted unless the regulatory system is reformed. Many believe that regulatory reform should be given the highest priority—that an incentive-based regulatory and policy framework is of paramount importance and should precede efforts to accelerate the commercialization of clean vehicles.

In an economically rational world, they would be correct. But they discount politics. Motorists vigorously oppose restrictions and taxes on their behavior unless they have an attractive alternative. Until a very clean vehicle is in the showroom, motorists will fight higher taxes on dirty gas guzzlers, even if those taxes are part of a revenue-neutral fee-bate. Thus, it is my firm conviction that proposals presented here to create more flexible and market-based regulations should be urgently pursued. And, if they are to work effectively, they must be coupled with the development and marketing of technologies such as those described in this book. In other words, efforts to reform regulation and accelerate technology development should go hand in hand. The advantages of pursuing both would far outstrip any advantage inherent in independent efforts.

The problem with the technology development side of the equation is not capital. The capital resources for developing new technologies in the United States are vast. In the first half of the 1990s, the auto and oil industries spent tens of billions of dollars to clean up gasoline and gasoline combustion. Those investments were made to preserve the status quo. The challenge to regulators and policymakers is simple: reduce uncertainty and provide incentives for industry so that it will channel its vast capital into the development and commercialization of more environmentally benign fuels and engines.

While the challenge may sound simple, the task of creating and administering innovative policies and regulations is not. The concept of a level playing field is nonsense. The corporate cultures of different industries and the attributes of new fuels and vehicles are so diverse that a perfectly fuel-neutral regulatory program could never be realized. And no program can completely eliminate abuses, such as large, politically motivated subsidies for corn-based ethanol. But it is possible to devise a more coherent and comprehensive framework that reduces the temptation for ad hoc political intervention at the first sighting of inequity or regulatory failure.

The key to undermining political favors for particular fuels or vehicle suppliers is the creation of a regulatory framework that sends strong, consistent signals to consumers and vehicle and fuel suppliers. Debate over fuels, engines, and vehicles would then enter a structured regulatory arena, making the excuse of flawed or imbalanced regulations less compelling. A more flexible and incentive-based framework must be the goal of regulatory reform.

But ideological purity has no place in the formulation of transportation and energy strategies, and it would be unwise to become overly enamored with the theory of market-based regulation. Flexibility and market incentives have their limits, too. There is still a need for selective use of command-and-control mandates. The ZEV mandate in California is a prime example. No imaginable set of incentives and subsidies could ever have achieved in such a short time what the ZEV mandate has achieved. In just a few years, it stimulated a surge of investment in batteries and electric vehicles, moving them from the backyard into the corporate boardroom.

CHAPTER 8

Technology Policy for Sustainable Transportation

Sustainability need not be painful. Ideally, it means exploiting technology for the social good. It can be attained through more diversity and more technology, giving us greater choice of vehicles and propulsion.

Sustainability cannot happen by itself. It requires leadership, not from the mainstream automobile industry or technology-based companies, but primarily from government. This is not to say that government is all-knowing or should be all-powerful. It is simply that government is better equipped than the market system to encourage industry and consumers to strive for less environmental damage, renewable energy, and livable communities. The government can redirect market signals. Only then will consumers and industry vigorously pursue benign and sustainable transportation.

A call for government leadership might have met strong resistance as recently as the 1980s. Then government's role was seen as primarily as-suring unfettered competition. That view was given pause by the dramatic success of government-orchestrated economic surges in countries such as Japan, Taiwan, South Korea, and Singapore.[1] Still, Americans remain deeply skeptical of centralized government, an attitude reinforced by economic disasters in the Soviet Union and other centrally planned economies of Eastern Europe. The result of these mixed experiences is a less ideological electorate that takes a more pragmatic attitude toward the opportunities and limitations of government activism.

What this book proposes is government leadership that allows consumers, industry, and society to embrace and pursue their own vision of the future. Government can encourage greater experimentation with vehicles and fuels, accelerating investments in those that are most successful. It can use its power to regulate and encourage changes in how vehicles are manufactured and used. In general, this means facilitating, not substituting for, private investment. It means unshackling the entrepreneurial spirit.

The story of two fuel cell bus programs illustrates the perils of excessive intervention.[2] Recall that in 1987 the U.S. Department of

Energy (DOE) set out to build a fuel cell bus. One bus was completed in April 1994, with two near-identical copies to be completed later. DOE spent at least $20.3 million over the seven-year period to design and construct its three buses, plus an undetermined additional amount for government contract managers, auditors, and other "support" personnel.

In late 1990, more than three years after the DOE bus project had begun, Ballard Technologies of Vancouver, Canada, set out on the same path. In March 1993, Ballard unveiled a fully operating bus for $3.8 million. The Ballard bus was built more quickly and cheaply. Why?

Part of the explanation is that the DOE bus was more complex and sophisticated. It had a hybridized system of fuel cells and batteries and used a reformer, while the Ballard bus ran strictly on hydrogen-fed fuel cells. Moreover, the DOE bus was designed for possible commercial service, with air-conditioning and long driving ranges, while the Ballard bus was built solely to demonstrate fuel cell technology. More telling, however, is that the DOE bus was managed by government, while the Ballard bus was managed by a private company with minimal government intervention.

The DOE project's high costs and long delays were the result of litigation, reams of paperwork, onerous regulations, problems coordinating hordes of contractors, and meddling. The woes began when the losing bidder contested the primary contract. This led to a delay of one and a half years. Further delays resulted from micromanagement as DOE contract personnel repeatedly challenged the contractors and project manager's technical decisions. Audits of contracts proceeded slowly, forcing contractors to stop work for months at a time.

Micromanagement created other problems as well. Specifications were imposed that could not be met by commercially available parts; this required special ordering, drove up costs, and slowed procurement. For instance, specified "radiation-hardened" computer chips were in short supply and could not be procured immediately or cheaply. Micromanagement also resulted in too many contractors, which had the effect of lessening commitment to the project. When faced with deadlines and problems, contractors felt less responsible than they might have otherwise.

The DOE's bus project illustrates not that government employees are incompetent, but rather that government is not suited to the development of technologies intended for commercial application.

In a broader sense, perhaps more troubling than excessive intervention is the fragmentation of government oversight and the current emphasis on short-term goals, particularly legislation that has as its binding criterion short-term air-quality improvement. Government has the potential to be more effective and farsighted. Listed below, and elaborated upon in the remainder of this chapter, are four recommendations for government initiatives that go beyond the current preoccupation with short-term air-pollution reduction.

1. Retain the ZEV mandate to jumpstart investments in electric propulsion;

2. Restructure and expand green-car research to support development of commercially risky, longer-term technologies such as fuel cells;

3. Reform the regulatory process to make it more flexible and to provide incentives for the introduction of environmentally benign vehicles and fuels; and

4. Support innovative efforts by local government to expand and redirect transportation choices.

The Zero-Emission Mandate[3]

The intent of the ZEV mandate that CARB adopted in 1990 is to accelerate the development and deployment of ZEV technologies. The goal is not necessarily 2 percent ZEVs in 1998—that percentage was chosen somewhat arbitrarily to spur automaker action—but rather a much larger percentage in future years. Because CARB has no authority outside air quality, the mandate is based solely on air-quality benefits.

Opposition from the Big Three

Automaker opposition to the mandate is rooted in economics, organizational inertia, and corporate philosophy. The economic argument is that battery technology is not yet commercially ready—that if sold now, electric cars would cost too much and have unacceptably short driving ranges. Carmakers argue that each electric vehicle (with batteries) will cost an extra $10,000 to $20,000 above what a comparable gasoline car would cost, and therefore, that the price of conventional cars would have to be raised by as much as $2,000 or more to compensate for that loss (the amount depending on how many electric vehicles were sold outside California and on the mandate level).[4] An option would be for auto companies to pay the $5,000 penalty, but major manufacturers would do that only as a last resort. Ford has publicly declared that it will adhere to whatever law is in place.

The automakers' opposition to the ZEV mandate has some merit—initial electric vehicles indeed will have to be sold at a loss—but it is greatly overstated. With normal manufacturing and engineering improvements, costs would diminish, and as sales expanded, economies of scale would expand. As automakers came to accept that limited-range vehicles would sell, batteries could be shrunk and costs would further diminish. And as more low-cost NEVs are sold, the smaller the economic losses. Moreover, manufacturers could minimize costs initially by inserting electric drivelines into existing gasoline vehicles. They could even sell engine-less gliders to smaller conversion companies and let them do the retrofits. Various companies have already announced plans to pursue one or more of these strategies.

In any case, the notion of one type of vehicle cross-subsidizing another is not unique. Large automakers typically sell small cars at little or no profit, or even a loss, subsidizing them with profits from larger cars and trucks. They do this to meet average fuel-economy standards and to hook young buyers, who will presumably move up to larger, more profitable cars later.

Another economic argument against the ZEV mandate is that air pollution could be reduced at less cost by other means. That argument is correct—in the short run. But as noted, the goal of the ZEV mandate is to accelerate the development and deployment of electric-drive technology for the long term, not just to improve air quality in the near future. Large reductions of emissions (including greenhouse gases) and energy use are not possible with ICE vehicles (with the possible exception of biomass-fuel vehicles). As ZEVs become less expensive and more widespread, pollution will be further reduced and at less cost. The automaker argument is that instead of bearing the brunt of large losses initially, they should be allowed to develop the technology in the laboratory. If one could elicit a promise that knowledge gained in the laboratory would be quickly transferred to the marketplace, then their request might be reasonable. But such promises are difficult to enforce.

Based on many interviews with executives, I sense that opposition to the mandate goes far beyond quarterly profit statements. At some level, Detroit executives recognize that their companies are culturally and organizationally unprepared for electric propulsion. They are clearly not prepared to manufacture the many new components needed for electric vehicles. But the corporate unease goes much deeper. There is a feeling that electric vehicles will succeed only if new ways are found to manufacture, sell, and service them—ways not compatible with the current structure and ingrained practices of today's industry. It may be that automakers are not agile enough to transform themselves in a timely manner—adopting new materials and components, collaborating with new manufacturing companies, reducing the size of production runs, designing new ways of retailing and servicing vehicles, and replacing a mechanical-engineering workplace with an electrical-engineering one. A vast cultural and organizational gulf exists between current industry practices and the practices that may be necessary for success in the future. The gulf opens up business opportunity for new entrants from outside the auto industry and outside Michigan. Automakers acknowledge this threat but, especially in the case of the Big Three, appear unprepared to respond.[5]

Today's major automakers also seem blind to changes in the consumer market and therefore to market-niche opportunities for electric vehicles. Detroit executives were typically raised in single-car households with a male head. They cling to an outdated 1950s image of the market. Perhaps still heavily influenced by the macho tilt of car-enthusiast magazines, they cannot fully appreciate the opportunities created by the emergence of women buyers and multicar households. No longer must all vehicles serve all purposes, and no longer must all vehicles have 300 miles of range. Auto executives remain almost oblivious to the value of home recharging, green cars, and low maintenance. They think in terms of the "typical" driver, not the potential 2–5 percent of buyers who would comprise the initial electric vehicle market niche.

Jumpstarting Innovation

The ZEV mandate has probably spurred more progress in electric propulsion technology in a few years than took place over the course of twenty years under the combined auspices of the auto industry and DOE. Almost entirely because of the mandate, every major automaker in the world has now invested in electric vehicle development. Hundreds of companies are sprouting up to develop and commercialize critical technologies of the future, such as flywheels, batteries, ultracapacitors, and fuel cells, as well as less critical technologies such as super-efficient car heaters and tires. These outsiders are given a boost by a little-appreciated feature of the ZEV mandate—ZEV credits are tradable. Thus mainline manufacturers not adept at building niche vehicles will purchase credits from smaller companies, providing them with cash payments.

It is important not to tinker with the basic thrust of the ZEV mandate until after 1998. Many companies are making substantial investments in future technologies and products based on that mandate. Any indication that the mandate might be changed or abandoned would freeze investments in hundreds of companies, especially in small companies dependent on outside financing.

In summary, the ZEV mandate is a blunt but effective instrument for overcoming market uncertainty, corporate resistance, and start-up barriers. Given the automakers' history of hindering efforts to reduce pollution and energy consumption in cars, government has little choice but to turn to such an instrument to jumpstart investments. In this case, corporate interests are not the public interest. Indeed, the ZEV mandate may prove to be the most pivotal event in automotive history since Henry Ford introduced mass production eighty years ago.

Green, Clean, and Super

If the momentum toward environmentally benign fuels and vehicles is to be sustained, investment in riskier, long-term technologies will have to be accelerated by means of a research partnership between government and industry. The end of the cold war has been a catalyst for such partnership. The U.S. government finds itself with a large number of national laboratories whose existence can no longer be justified on military and security grounds. One solution is a guns-to-butter conversion of these facilities to civilian applications, especially advanced technologies for transportation.

The Partnership for a New Generation Vehicle
This, indeed, was the essence of an initiative announced on September 29, 1993, by President Clinton. Initially known as the Clean Car Initiative (and as the Supercar Initiative by the automotive industry), its formal name is now the Partnership for a New Generation Vehicle. It is intended to accelerate the development of electric propulsion, ultra-efficient technologies, lightweight materials, and advanced manufacturing processes. An undetermined amount of funding will be diverted at the labs to advanced transportation technologies—with no new funding in the works—and lab personnel and equipment will be reassigned to work with the Big Three and their suppliers.

In theory, everyone benefits. The labs would have a renewed mission, thousands of highly trained scientists and engineers would be kept productively employed, and automakers would receive a much-needed infusion of technical know-how. Advanced technology developed during the well-funded days of the cold war would be transferred to the civilian economy. It would invigorate the beleaguered national laboratory system and speed the transition to more environmentally benign technologies. The Partnership for a New Generation Vehicle is touted as a modern counterpart to the 1960s Apollo effort to put a person on the moon.

The laboratories do have a storehouse of knowledge and technology that could be of immediate value to commercial enterprises. While one might question the procedures used in making this information available—"cooperative research and development agreements" are negotiated on a first-come, first-serve basis and contain exclusivity clauses—the net effect on the economy will undoubtedly be positive.[6]

But how much so? One cause for skepticism is the absence of new funding, though this is not necessarily a critical flaw. Over $3 billion per year is spent at three large national weapons laboratories (Lawrence Livermore, Los Alamos, and Sandia) and almost another $3 billion is spent at six other multipurpose energy laboratories (Argonne, Oak Ridge, Idaho, Brookhaven, Lawrence Berkeley, and Pacific Northwest), with almost $20 billion more spent at hundreds of small federally owned or funded national laboratories.[7] If even a small fraction of these budgets were (productively) diverted to vehicle and propulsion technologies, their performance and cost would improve dramatically and quickly.

A second and more substantial concern is whether laboratories, especially weapons laboratories, are suited for the new research. On the positive side, the talent and equipment in such facilities are among the finest in the nation.[8] While expertise in weapon design and other exotica is not of use to the automotive and fuel industries, the knowledge and technologies embedded in that expertise clearly are. The principal contribution of weapons laboratories to industry will be fundamental technological innovation.

But weapons laboratories were not intended to design products destined for the marketplace. Their strength is basic science and designing high-performance equipment, not designing products and devising manufacturing processes for mass production and low cost. If basic science were industry's critical missing ingredient, then these laboratories might have an important role to play. That is not the case.

Many promising technological concepts already exist in prototypes, including fuel cells, ultracapacitors, flywheels, batteries, and lightweight materials. Industry's principal need over the next ten to fifteen years is not new science or technology, but improved and cheap technology. This is an engineering and manufacturing challenge with which weapons laboratories have only passing acquaintance. It may be a national priority to save the jobs and expertise of these institutions, but that goal appears only tangentially compatible with the goal of making cleaner and more-energy-efficient vehicles.

A third, particularly troubling concern about the Partnership for a New Generation Vehicle is its focus on the Big Three. Those companies certainly need to be intimately involved in the initiative, but not as the dominant players. Perhaps it is appropriate to allow them to direct the research agenda for incremental technologies, such as some materials and manufacturing processes, of which they and their suppliers will be the principal and immediate users. However, they probably should not play a central role in advanced propulsion technologies, given their ambivalence to commercializing electric-drive technologies.

Government should be actively seeking out other technology companies and research organizations and forging closer links between them and the national laboratories. These other organizational companies will be more aggressive in advancing the technology, and the non-auto companies more aggressive at commercializing it. If nothing else, these other efforts will provide a benchmark for assessing automaker progress. It would take substantial effort to involve more companies and universities, but it likely will result in quicker commercialization of research.

A fourth concern is the partnership initiative's stated purpose: to advance technology to the point where government regulation is no longer necessary—that is, according to a September 1993 press release, to "replace lawyers with engineers." While some high-level officials have suggested in private that this and other such statements were pure rhetoric to satisfy the automakers, the professed goal undermines the initiative's credibility. So long as automakers feel no regulatory pressure to commercialize their research, the movement of technology from the lab to the marketplace will be slow.

Improving on the Partnership

Government-industry partnerships of the type being forged could indeed accelerate the development of advanced propulsion and vehicle technologies. The most productive partnerships would do the following: expand well beyond weapons labs and major automakers to include smaller labs and technology companies; place limits on proprietary agreements between national labs and private companies when public funds are involved; provide new, not just diverted, funding; require commercialization agreements of manufacturers (or stiffened regulatory standards) in return for government funding; and create small, independent research facilities at universities and perhaps elsewhere as a benchmark of what is possible and to evaluate the progress of the partnership.

Technology Plus Regulatory Reform

All things being equal, sustained change is more effectively achieved by harnessing market forces than by imposing regulations. The exception to the rule is when start-up costs are very large, as with electric propulsion—and therein lies the justification for the ZEV mandate. The ZEV mandate, however, cannot stand alone, nor should it stand long. It must be supplemented and eventually supplanted by broader initiatives aimed at altering consumer and industry behavior. Ideally, future initiatives would encourage transportation choices that reduced social cost, would take into account more than just air quality, and would rely as much as possible on market forces.

Regulatory and demand-altering initiatives would have little effect by themselves, and more often than not they would be rejected in the political arena unless coordinated and combined with technology initiatives. The benefits of coordinating technology with policy and regulatory initiatives are potentially huge. Government should be creating mechanisms such as taxes, tax credits, fees, and marketable credits that marry technology and regulatory initiatives. This would lead to a more flexible, incentive-based public policy and also create the framework needed to guide business and consumer decisions toward a sustainable future. Such an approach would not only be more effective, it would also be more appealing politically, as indeed it must be.

Technology as an Enabler

Those who argue that government should focus on better pricing of fuels and vehicles, rather than support technology initiatives, are missing an important point. The political reality is that taxing vehicles and fuels faces formidable opposition from drivers who see no alternative to higher prices and from an industry that sees shrinking markets and revenues. If consumers and industry felt that they had attractive alternatives, or if the new taxes and other restrictions were paired with new fuels, vehicles, and travel options, opposition would probably fade.

In light of this, technologies should be promoted not only for their direct energy and environmental benefits, but also as enablers of pricing and regulatory initiatives. For instance, if cleaner fuels were made available, it would be much easier to pass taxes on dirty fuels and cars. Likewise, demonstrating the viability of hybrid and fuel cell technology would allow for stiffer taxes on emissions and energy use. Providing choices to consumers and manufacturers makes new taxes and more stringent standards more palatable by lifting the aura of punishment.

Consider recent experiences with emission standards. During the 1980s, efforts to strengthen emission standards on both light-duty gasoline and heavy-duty diesel engines were stymied by industry. It was not until cleaner-burning natural gas and methanol fuels emerged as viable alternatives in the latter part of the decade that regulators were emboldened to break the deadlock. As it turned out, industry responded to the new standards by redoubling their efforts to reduce emissions from gasoline and diesel fuel, successfully so. While it now appears that manufacturers will meet these stringent standards without switching to natural gas or methanol, the point is that the introduction of alternative fuels, even in a tentative manner, enabled the adoption of tighter standards.

One can imagine ZEV technology playing the same role in the future that alternative fuels played in the late 1980s. Advancements in this area might, for example, enable the adoption of more stringent emission and CAFE standards and of stiffer taxes on dirtier fuels and vehicles. One can imagine, in a conservative scenario, that electric-vehicle technology might be a weak presence in the market but that would serve as a kind of straw man for altering market signals and strengthening regulations; or, in a more adventurous scenario, that the altered signals and strengthened rules would dramatically accelerate the market penetration of electric vehicles.

Encouraging Innovative Communities

This last set of recommendations is aimed at making local communities more hospitable to small vehicles (and smart paratransit), and enabling more far-reaching changes in land use and mobility patterns. Small vehicles, especially when powered by electricity, could be an important component of a sustainable transportation system. They are less obtrusive and less environmentally damaging than larger vehicles, and for certain purposes they can provide virtually the same quality service at much less cost and using much less energy. They are also an ideal application of battery-powered electric propulsion.

The motivation for introducing neighborhood cars goes beyond direct environmental benefits. Accessibility and mobility could be increased by small station cars that connect with conventional transit services, and by partially automated neighborhood vehicles that elderly and physically impaired people can drive easily.

An example of a thwarted golden opportunity is Sutter Bay, a town designed in the late 1980s and early 1990s to house more than 100,000 people in an area north of Sacramento, California.[9] A group of progressive developers sought to create an environmentally sound community by incorporating the most sophisticated transportation, energy, and telecommunications technology available. They were planning to wire the community with fiber-optic cable for interactive television, teleconferencing, and other purposes. They also wanted to integrate various "smart" transportation technologies, including paratransit.

Perhaps the most innovative feature was the design of the urban road system around neighborhood electric vehicles (NEVs). They intended to provide a free NEV with the purchase of each house. The central means of transport within the community was to be walking, bicycling, and using NEVs, with conventional cars allowed only on a few roads and with parking behind houses.

Although the project was approved by the Sutter County government, a conservative antigrowth element pushed passage of a countywide referendum that forced a severe scaling back of the plan. Opponents included politicians who feared a geographical power shift from the more populated northern parts of the county to the rural south, conservative groups who predicted an invasion of undesirable outsiders, and residents who believed the rural ambiance of the area would disappear. Endorsements of the project by environmental groups served only to further antagonize residents. As this book goes to press, the destiny of the project has passed into the hands of the court system.

Sutter Bay is a far too common story of narrow self-interest and limited imagination overwhelming the greater public interest. Sutter Bay and developments like it can serve as a unique forum for testing and promoting new transportation concepts and technologies. With stronger support from government and business, such innovative towns would have a greater chance of success. Government, business, and academia need to take a more active stance in supporting efforts like these in all communities. Without that leadership, the status quo will prevail, and the opportunity to explore new concepts and technologies will be lost.

Under any conditions, the introduction of NEVs would proceed slowly. Consumer expectations would have to be changed, new traffic rules and laws adopted, new vehicle types manufactured. But perhaps the largest barrier is the absence of local leadership. Local governments in existing communities do not have the resources and expertise to orchestrate the introduction of small vehicles. Yet it is they who would oversee and manage the deployment of a recharging infrastructure, modify parking and traffic rules to accommodate station cars and NEVs, and create local incentives and rules for electric vehicles. Local governments need encouragement and help. That encouragement is not coming from air quality regulators or supporters of the ZEV mandate; they are more focused on demonstrating the viability of battery technology in full-size cars. Neither is it coming from automakers, who shun small vehicles, from state and national transportation departments, which have little expertise in vehicles, nor from safety regulators, who are skeptical of all small vehicles.

———————————

In closing, I note that many argue that the problems created by vehicle proliferation can be managed through small adaptations, such as reformulated gasoline, renewed emphasis on controlling emissions from gasoline combustion, more efficient techniques for handling internal combustion, and perhaps some use of natural gas and methanol. This incrementalist thinking does not, however, take into account such factors as accelerated climate change, dependence on oil from the precarious Persian Gulf, and increased travel due to expanding suburbs.

Do we want to accept air pollution as a permanent condition—sickening millions of people, damaging crops and buildings, and casting an ugly pall over our cities and parks? Are we to have faith that traffic congestion will be eased by new investments in transit, incremental improvements in smart-car technologies, and various demand-management programs? If the Persian Gulf were to deteriorate politically, or if emissions-induced climate change were to accelerate, future generations would face a deteriorating quality of life on all fronts—economic, political, and environmental. They would have to struggle to undo what was wrought by the wishful thinking of our generation. That is not a legacy I would like to leave.

Large positive changes are possible; indeed, they are at hand. We are on the verge of a technological revolution. We are also on the verge of a regulatory revolution. Let us nurture and exploit these opportunities. And let us proceed with some haste. Not to do so would be irresponsible.

Notes

Preface

1. The expensive technological solutions I offer in this book are largely irrelevant to the less affluent countries of the world. Those nations need an entirely new, and less expensive, ethos of transportation—a new model and vision. I came to this conclusion in part after years in Latin America and Asia as a Peace Corps planner and occasional consultant.
2. Elmer W. Johnson, "Taming the Car and Its User: Should We Do Both?" *Aspen Quarterly* (Autumn 1992). Based on a presentation by J. Meyer, Harvard University.
3. The product of that study was *Avoiding the Collision of Cities and Cars: Urban Transportation Policy for the Twenty-first Century* (Chicago: American Academy of Arts and Sciences, 1993).
4. Note that I do not use *sustainability* here. Because the word evokes the concepts and values I embrace, I included it in the book's subtitle. But if it is evocative it is also vague, and for that reason I have not used it much elsewhere in the book.
5. Aside from some insights in Chapters 1 and 2, I leave it to others to address the problem of traffic congestion, the consequence of ever-increasing travel. By advocating the use of smaller (electric) vehicles and smart paratransit, I go part of the way toward mitigating if not solving the problem of severely crowded roadways.

Chapter 1

1. G. Speth, in Steve Nadis and James J. Mackenzie, *Car Trouble* (Boston: Beacon Press, 1993).
2. Throughout this book, unless otherwise noted, *car* and *auto* refer to all light-duty vehicles, including pickups, vans, and other light-duty trucks. In most countries, cars far outnumber light-duty trucks, but in the United States light trucks have been gaining market share since the 1970s; by 1994, they accounted for almost 40 percent of light-vehicle sales. Most light trucks in the United States are used for personal transportation.
3. One important caveat is in order. While rising auto use is a global phenomenon that affects both rich and poor countries, the solutions are not equally relevant to both. Wealthier nations can allocate substantial resources to countering the ill effects of the automobile; poorer countries cannot. The auto-highway development model of North America and Western Europe requires not only taming the auto, but also the much more expensive proposition of building the vast network of highways to accommodate more and more vehicles. The poorest countries will have to follow a very different path, one based primarily on mass transit.

4. Elmer W. Johnson, *Avoiding the Collision of Cities and Cars: Urban Transportation Policy for the Twenty-First Century* (Chicago: American Academy of Arts and Sciences, 1993), p. 44

5. Ibid., p. v.

6. James H. Kunstler, *The Geography of Nowhere* (New York: Simon and Schuster, 1993); and Joel Garreau, *Edge City: Life on the New Frontier* (New York: Doubleday, 1991).

7. Elmer W. Johnson, "Taming the Car and Its User: Should We Do Both?" *Aspen Quarterly* (Autumn 1992). Statistics are based on a presentation by J. Meyer of Harvard University. See also Stephen Bernow and Mark Fulmer, "A Social Cost Analysis of Alternative Fuels for Light Duty Vehicles," in D. Sperling and S. Shaheen, eds., *Energy Strategies for a Sustainable Transportation System* (tentative title) (Washington, D.C., and Berkeley, California: American Council for an Energy-Efficient Economy, forthcoming); and Brian Ketcham, "Making Transportation Choices Based on Real Costs" (paper presented at the Transportation 2000 Conference, Snowmass, Colorado, October 6, 1991).

8. U. S. Congress, Office of Technology Assessment (OTA), "Saving Energy in U.S. Transportation,"OTA-ETI-S89 (Washington, D.C.: U.S. Government Printing Office, 1994). The analysis of social costs is based primarily on a larger report submitted to OTA by Mark A. Delucchi, "Annualized Social Costs of Motor Vehicle Use Based on 1990–91 Data" (Institute of Transportation Studies, University of California, Davis, 1994).

9. James Mackenzie, Roger Dower, and Donald Chen, *The Going Rate: What It Really Costs To Drive* (Washington, D.C.: World Resources Institute, 1992). See also Bernow and Fulmer, "Social Cost Analysis."

10. Delucchi, "Annualized Social Costs."

11. Michael Quanlu Wang, Danilo Santini, and Sonja Warinner, "Methods of Evaluating Air Pollution and Estimating Monetary Values of Air Pollutants" (Argonne National Laboratory, Argonne, Illinois, 1994).

12. The estimate of $2.50 was calculated by converting stationary-source control costs in dollars per ton of pollution removed to equivalent emissions from cars with average (lifetime) emissions and fuel economy sold in the mid-1990s. Marginal pollution-control costs in the early 1990s in California are about $26,000 per ton of nitrogen oxides and $19,000 per ton of reactive organic gases. Good estimates of carbon-monoxide costs from stationary sources are not available because most carbon monoxide comes from cars; based on other empirical studies, it was assumed that carbon-monoxide control costs per gallon are about the same as control costs per gallon of nitrogen oxides (taking into consideration that carbon monoxide emissions are about ten times greater per mile than nitrogen oxide emissions, but that control costs per ton are about ten times less).

13. Neil Postman, *Technology: The Surrender of Culture to Technology* (New York: Alfred A. Knopf, 1992); Wolfgang Zuckerman, *End of the Road: From World Crisis to Sustainable Transportation* (Post Mills, Vermont: Chelsea Green, 1991); Lester Brown, Christopher Flavin, and Colin Norman, *Running on Empty* (New York: W. W. Norton, 1979); Emin Tengström, *The Use of the Automo-*

bile: Its Implications for Man, Society, and the Environment (Stockholm: Swedish Transport Research Board, 1992).

14. While telecommunications is a promising substitute for commuting, shopping, and other trips, and should be encouraged, the need for supervision and personal interaction in the workplace, and the lack of suitable work space in many homes, effectively limits the potential for telecommuting. See Patricia Mokhtarian, Susan Handy, and Ilan Salmon, "Methodological Issues in the Estimation of Travel, Energy, and Air Quality Impacts of Telecommuting," *Transportation Research* A (forthcoming).

15. Michael Cameron, *Efficiency and Fairness on the Road: Strategies for Unsnarling Traffic in Southern California* (Oakland, California, Environmental Defense Fund, 1994).

16. Stephen Meyer, "Transportation in the LDCs: A Major Area of Growth in World Oil Demand" (Lawrence Berkeley Laboratory, Berkeley, California, March 1988, LBL-24198); Joel Darmstadter and Andrew Jones, "Prospects for Reduced CO_2 Emissions in Automotive Transport" (Resources for the Future, Washington, D.C., 1990, ENR 90-15).

17. Updated from Lee Schipper, Ruth Steiner, and Stephen Meyer, "Trends in Transportation Energy Use, 1970–1988: An International Perspective," in D. L. Greene and D. J. Santini, eds., *Transportation and Global Climate Change* (Washington, D.C., and Berkeley, California: American Council for an Energy-Efficient Economy, 1993), pp. 51–89. The diminishing role of autos in the United States is due to the increase in air travel, which accounted for 10 percent of all U.S. passenger travel by 1990.

18. General Accounting Office, "Reducing Vehicle Emissions With Transportation Control Measures" (GPO, Washington, D.C., 1993, GAO/RCED-93-169).

19. Anthony Downs, *Stuck in Traffic: Coping with Peak-Hour Traffic Congestion* (Washington, D.C.: Brookings Institution and Cambridge, Massachusetts: Lincoln Institute of Land Policy, 1992); Genevieve Giuliano, "Transportation Demand Management: Promise or Panacea?" *APA Journal* 58 (3): 327–35 (1992); David Hartgen and Mark A. Casey, "Television Media Campaigns to Encourage Changes in Urban Travel Behavior: A Case Study," *Transportation Research Record* 1285: 30–39 (1990).

20. G. Giuliano, K. Hwang, and M. Wachs, "Evaluation of a Mandatory Transportation Demand Management Program in Southern California," *Transportation Research* 27A(2): 125–38.

21. South Coast Air Quality Management District, *Air Quality Digest,* Diamond Bar, California, Winter 1988.

22. Based on estimates by Lamont Hempel in a draft paper for the conference "IVHS and the Environment," sponsored by IVHS America, Washington, D.C.; and G. Guliano and M. Wachs, "Managing Transportation Demand: Markets versus Mandates" (Reason Foundation paper 148, Los Angeles, 1992). The $3,000 in costs includes cash incentives given to workers and salary and benefits of staff to manage the program.

23. Michael Cameron, *Transportation Efficiency: Tackling Southern California's Air*

Pollution and Congestion (Oakland: Environmental Defense Fund, 1991). Analysis was conducted by Greig Harvey.

24. Martin Wachs, "U.S. Transit Subsidy Policy: In Need of Reform," *Science* 244: 1545–49 (1989).

25. Oak Ridge National Laboratory, *Transportation Energy Data Book*, 13th ed. (Springfield, Virginia: NTIS, 1993, ORNL-6743, pp. 4–12); U.S. Federal Highway Administration, *1990 Nationwide Personal Transportation Study: Summary of Travel Trends* (Washington, D.C.: GPO, March 1992).

26. Schipper *et al.*, "Trends in Transportation Energy Use."

27. John Pucher, "Urban Travel Behavior as the Outcome of Public Policy: The Example of Modal-Split in Western Europe and North America," *Journal of the American Planning Association* 54: 509–20 (Autumn 1988).

28. Oak Ridge National Laboratory, *Transportation Energy Data Book*, 14th ed. (Springfield, Virginia: NTIS, 1994).

29. Thomas Turrentine, Daniel Sperling, and Kenneth Kurani, "Market Potential of Electric and Natural Gas Vehicles" (report RR-92-8, University of California, Davis, Institute of Transportation Studies, 1992), p. 20.

30. Q. Wang, D. Sperling, and C. Kling, "Light Duty Vehicle Exhaust Emission Control Cost Estimates Using a Part-Pricing Approach," *Journal of the Air and Waste Management Association* 43: 1461–71 (1993).

31. National Research Council, *Automotive Fuel Economy* (Washington, D.C.: National Academy Press, 1992).

Chapter 2

1. See, for example, Doug Lawson, "Passing the Test: Human Behavior and California's Smog Check Program," *Journal of the Air and Waste Management Association* 43 (1993); and Amihai Glazer, Daniel Klein, and Charles Lave, "Clean for a Day: Troubles with California's Smog Check" (report to the California State Senate Committee on Transportation, University of California Transportation Center, Berkeley, August 1993).

2. J. G. Calvert, J. B. Heywood, R. F. Sawyer, *et al.*, "Achieving Acceptable Air Quality: Some Reflections on Controlling Vehicle Emissions," *Science* 261: 37–45 (2 July 1993); Buzz Breedlove, *Motor Vehicle Inspection and Maintenance in California* (Sacramento: California Research Bureau, 1993).

3. Charles Lave, "Clean for a Day: California versus the EPA's Smog Check Mandates," *Access* magazine, University of California Transportation Center, Berkeley, 1993.

4. U.S. Congress, Office of Technology Assessment, *Retiring Old Cars: Programs to Save Gasoline and Reduce Emissions* (Washington, D.C.: Government Printing Office, 1992).

5. A. Albertini, D. Edelstein, W. Harrington, et al., "Reducing Emissions from Old Cars: The Economics of the Delaware Vehicle Retirement Program" (Resources for the Future Discussion Paper 94-27, Washington, D.C., 1994); Illinois Environmental Protection Agency, "Pilot Project for Vehicle Scrapping in Illinois," (IEPA, Springfield, Illinois, 1993); Unocal Corp., "Unocal

SCRAP: A Clean-Air Initiative from Unocal" (Unocal, Los Angeles, California, 1991).

6. Shi Ling Hsu and D. Sperling, "The Uncertain Air Quality Impacts of Automobile Retirement Programs," *Transportation Research Record* (forthcoming).

7. Many news reports and even scholarly publications claim that vehicle emissions have been reduced by 96 to 99 percent. This is not true. The supposed improvements are calculated by comparing current standards for hydrocarbon and carbon-monoxide emissions against actual emissions from cars in the early 1960s (before controls were imposed). The flaw in this calculation is that the emissions measured in the 1960s were probably fairly accurate, while those of today are grossly inaccurate—underestimating emissions by a factor of two to four in the case of hydrocarbons and carbon monoxide (see National Research Council, *Rethinking the Ozone Problem in Urban and Regional Air Pollution* [Washington, D.C.: National Academy Press, 1991]. This apparent anomaly in testing accuracy has to do with changes in engine and emission technology.

8. Q. Wang, D. Sperling, and C. Kling, "Light Duty Vehicle Exhaust Emission Control Cost Estimates Using a Part-Pricing Approach," *Journal of the Air and Waste Management Association* 143: 1461–71 (1993). It is impossible to document emission control costs because many components perform multiple functions and automakers will not assist independent analysts.

9. David L. Greene, "CAFE or Price: An Analysis of the Federal Fuel Economy Regulations and Gasoline Prices on New Car MPG, 1978–89," *Energy Journal* 11(3): 37–57 (1990).

10. R. A. Leone and T. W. Parkinson, "Conserving Energy: Is There a Better Way? A Study of Corporate Average Fuel Economy" (paper prepared for the Association of International Automobile Manufacturers, May 1990).

11. C. Difiglio, K. G. Duleep, and D. L. Greene, "Cost Effectiveness of Future Fuel Economy Improvements," *Energy Journal* 11 (1): 65–86 (1990); U.S. Congress, Office of Technology Assessment, *Improving Automobile Fuel Economy: New Standards, New Approaches* (Washington D.C.: Government Printing Office, 1991); National Research Council, *Automotive Fuel Economy: How Far Should We Go?* (Washington, D.C.: National Academy Press, 1992).

12. IVHS America, "National Program Plan" (Washington, D.C.: IVHS America, 1993).

13. Randolph W. Hall, "Non-Recurrent Congestion: How Big Is the Problem? Are Traveler Information Systems the Solution?" *Transportation Research* 1C (1): 89–103 (1993); Haitham Al-Deek, Michael Martello, Adolph D. May, *et al.,* "Potential Benefits of In-Vehicle Information Systems in a Real-Life Freeway Corridor under Recurring and Incident-Induced Congestion," *Proceedings of the Vehicle Navigation and Information Systems Conference* (Toronto and New York: IEEE, 1989), pp. 288–91.

14. D. C. LeBlanc, M. Meyer, M. Saunders, *et al.,* "Carbon Monoxide Emissions from Road Driving: Evidence of Emissions Due to Power Enrichment," *Transportation Research Record* (forthcoming).

15. For one example of emission benefits from smoothing traffic, see Simon Washington and Randall Guensler, "Carbon Monoxide Impacts of Automatic Vehicle Identification Applied to Vehicle Tolling Operations," *Proceedings of the Conference on Intelligent Transportation and the Environment* (Washington, D.C.: IVHS America, forthcoming).

16. R. A. Johnston, M. DeLuchi, D. Sperling, *et al.,* "Automating Urban Freeways: Policy Research Agenda," *Journal of Transportation Engineering* 116 (4): 442–60 (1990).

17. Ibid.

18. See *Proceedings of the Conference on Intelligent Transportation and the Environment* (Washington, D.C.: IVHS America, forthcoming).

19. Robert Johnston and Raju Ceerla, "A Systems-Level Evaluation of Automated Urban Freeways: Effects on Travel, Emissions, and Costs," *Journal of Transportation Engineering* (forthcoming).

20. Deborah Gordon, "Intelligent Vehicle/Highway Systems: An Environmental Perspective," in Jonathan L. Gifford, T. A. Horan, and Daniel Sperling, eds., *Transportation, Information, and Public Policy* (Davis, California: Institute of Transportation Studies, 1992), pp. 9–28; David Burwell, "Is Anybody Listening to the Customer?" *IVHS Review* (Summer 1993): 17–26.

21. Daniel Sperling and Michael Replogle, "IVHS and the Environment," *Proceedings of the Conference on IVHS and the Environment* (Washington, D.C.: IVHS America, forthcoming).

22. For example, see J. Stone, A. Nalevanko, and G. Gilbert, "Computer Dispatch and Scheduling for Paratransit: An Application of Advanced Public Transportation Systems," *Transportation Quarterly* 48 (2): 173–84 (Spring 1994); and Robert Cervero, "Fostering Commercial Transit: Alternatives in Greater Los Angeles," *Policy Insight* (Reason Foundation, No. 146, Los Angeles, 1992), 34 pp.

23. Michael Cameron, *Efficiency and Fairness on the Road: Strategies for Unsnarling Traffic in Southern California* (Oakland: Environmental Defense Fund, 1994).

24. Daniel Sperling, *New Transportation Fuels: A Strategic Approach to Technological Change* (Berkeley: University of California Press, 1988).

25. Synthetic Fuel Corporation, "Comprehensive Strategy Report" (SFC, Washington, D.C., 1985).

26. National Research Council, *Rethinking the Ozone Problem in Urban and Regional Air Pollution* (Washington, D.C.: National Academy Press, 1991); Auto/Oil Air Quality Improvement Research Program, "Air Quality Modeling Results for Reformulated Gasolines in Years 2005/2010," *Technical Bulletin* 3 (May 1991); Auto/Oil Air Quality Improvement Research Program, "Emissions of Toxic Air Pollutants Using Reformulated Gasolines," *Technical Bulletin* 5 (June 1991); D. E. Gushee, "Alternative Fuels for Automobiles: Are They Cleaner Than Gasoline?" (Washington, D.C.: Congressional Research Service, Library of Congress, 92-235 S, 1992); U.S. General Accounting Office, "Gasoline Marketing: Uncertainties Surround Reformulated Gasoline as a Motor Fuel" (Washington, D.C., 1990, GAO/RCED-90-153).

27. National Research Council, *Rethinking the Ozone Problem;* Gushee, "Alternative Fuels for Automobiles"; Calvert *et al.*, "Achieving Acceptable Air Quality"; A. G. Russell, D. St. Pierre, and J. B. Milford. "Ozone Control and Methanol Fuel Use," *Science* 247: 201–5 (1990); A. J. Krupnick and M. A. Walls, "The Cost-Effectiveness of Methanol for Reducing Motor Vehicle Emissions and Urban Ozone," *Journal of Policy Analysis and Management* 11 (3), 373–96 (1992).

28. National Research Council, *Rethinking the Ozone Problem,* p. 400.

29. R. W. Hahn, "The Economics of Methanol," Auto/Oil Air Quality Improvement Research Program, *Economics Bulletin* 1 (1992); National Research Council, *Fuels to Drive Our Future* (Washington, D.C.: National Academy Press, 1990).

30. Sperling, *New Transportation Fuels;* Ryan E. Katofsky, "The Production of Fluid Fuels from Biomass" (Ph.D. diss., Princeton, University, Center for Energy and Environmental Studies, 1993).

31. Sperling, *New Transportation Fuels.* Brazil is a world leader in designing and optimizing the operation of distillation plants, using the bagasse byproduct for fuel.

32. *Highway Fact Book, 1992* (Washington, D.C.: Highway Users Federation, 1992).

33. Calvert *et al.*, "Achieving Acceptable Air Quality"; Office of Technology Assessment, *Replacing Gasoline: Alternative Fuels for Light-Duty Vehicles* (Washington, D.C.: GPO, 1990).

34. Mark A. DeLuchi, "Emissions of Greenhouse Gases from the Use of Transportation Fuels and Electricity" (Argonne National Laboratory, Center for Transportation Research, Argonne, Illinois, 1991, ANL/ESD/TM-22). For summary of findings, see M. DeLuchi, "Greenhouse-Gas Emissions from the Use of New Fuels for Transportation and Electricity," *Transportation Research* 27A (3): 187–91 (1993).

35. Eric D. Larson, "Technology for Electricity and Fuels from Biomass," *Annual Review of Energy and the Environment* 18: 567–630 (1993).

36. J. H. Cook, J. Beyea, and K. H. Keeler, "Potential Impacts of Biomass Production in the United States on Biological Diversity," *Annual Review of Energy and the Environment* 16: 401–31 (1991).

37. D. O. Hall, H. E. Mynick, and R. H. Williams, "Alternative Roles for Biomass in Coping with Greenhouse Warming," *Science and Global Security* 2: 113–51 (1991).

38. D. Sperling and M. A. DeLuchi, "Transportation Energy Futures," *Annual Review of Energy and the Environment* 14: 375–424 (1989); *Choosing an Alternative Fuel: Air Pollution and Greenhouse Gas Impacts* (Paris: OECD, 1993).

39. J. A. Alson, J. M. Adler, and T. M. Baines, "Motor Vehicle Emission Characteristics and Air Quality Impacts," in *Alternative Transportation Fuels: An Environmental and Energy Solution,* edited by D. Sperling (Westport, Connecticut: Quorum Books, 1989), pp. 109–44.

40. Southern California Association of Governments, "Highway Electrification and Automation Technologies—Regional Impacts Analysis Project (University of California, Berkeley, PATH, 1993).

41. K. Nesbitt, D. Sperling, and M. DeLuchi, "An Initial Assessment of Roadway-Powered Electric Vehicles," *Transportation Research Record* 1276: 41–55 (1990).

Chapter 3

1. Dean Drake of General Motors claims that 38 percent of cars were electric vehicles in 1900, 40 percent were steam powered, and 22 percent were gasoline powered (see Dean Drake, "Technology, Economics, and the ZEV Mandate," in D. Sperling and S. Shaheen, eds., *Energy Strategies for a Sustainable Transportation System* [tentative title] [Washington, D.C., and Berkeley, California: American Council for an Energy-Efficient Economy, forthcoming]). Ernest Wakefield, in his history of electric vehicles, is unable to provide total market statistics but does cite 1903 vehicle registration figures for New York State that indicate electric cars accounted for 20 percent of vehicles. Steam-powered cars accounted for 53 percent and gasoline-powered cars 27 percent (Ernest Wakefield, *History of the Electric Automobile* [Warrendale, Pennsylvania: Society of Automotive Engineers, 1994]).

2. *Motor* magazine (April 1904), cited in Virginia Scharff, *Taking the Wheel: Women and the Coming of the Motor Age* (New York: The Free Press, 1991).

3. Cited in Drake, "Technology, Economics, and the ZEV Mandate."

4. Scharff, *Taking the Wheel.* For additional history of electric vehicles, see Michael B. Schiffer, *The Road Not Taken: The Electric Automobile in America* (tentative title) (Washington, D.C.: Smithsonian Institute Press, forthcoming) and Wakefield, *History of the Electric Automobile.*

5. *Motor* magazine (1915), cited in Scharff, *Taking the Wheel.* Statements elsewhere suggest there were 34,000 electric vehicles on the road in 1912 (see Ken Ruddock, "Recharging an Old Idea: The Hundred-Year History of Electric Cars," *Automotive Quarterly* 31 (1) [1992]).

6. Wakefield, *History of the Electric Automobile.*

7. For detailed accounts, see ibid.

8. Ibid., p. 337.

9. *U.S. Petroleum Refining: Meeting Requirements for Cleaner Fuels and Refineries* (Washington, D.C.: National Petroleum Council, 1993).

10. David Sedgwick and Bryan Gruley, "You Guys Aren't Going to Make Us Build That Car, Are You?" *Detroit News and Free Press,* May 8, 1994.

11. Cited in James MacKenzie, *The Keys to the Car: Electric and Hydrogen Vehicles for the 21st Century* (Washington, D.C.: World Resources Institute, 1994), p. 35.

12. Auto manufacturers said adoption of the ZEV mandate and other low-emission vehicle rules by other states would in effect create a prohibited additional standard that was neither California's nor the federal one. They argued that the EVs (and other low-emission vehicles) would have to be different from those in California, in part due to the effect of cold weather on EV batteries.

13. Cited in Sedgwick and Gruley, "You Guys Aren't Going to Make Us Build That Car?"

14. Diana Kurylko, "French EVs Hinted for California," *Automotive News* (March 14, 1994): 1.

15. H. Dowlatabadi, A. J. Krupnick, and A. Russell, "Electric Vehicles and the Environment: Consequences for Emissions and Air Quality in Los Angeles and U.S. Regions" (Washington, D.C., Resources for the Future Discussion Paper QE91-01, October 1990); Northeast States for Coordinated Air Use Management (NESCAUM), "Impact of Battery-Powered Electric Vehicles on Air Quality in the Northeast States" (report prepared by Michael Tennis, Boston, July 1992); Q. Wang, M. A. DeLuchi, and D. Sperling, "Emission Impacts of Electric Vehicles," *Journal of the Air and Waste Management Association* 40: 1275–84 (1990).

16. Environmental Protection Agency, "Preliminary Electric Vehicle Emissions Assessment" (draft report, Washington, D.C., November 1993). The EPA later disowned the results of this study, saying that it was intended only to develop a methodology. As of August 1994, a final report had still not been issued. See Sierra Research and Charles River Associates, "The Cost-Effectiveness of Further Regulating Mobile Source Emissions" (report prepared for the American Automobile Manufacturers Association, Sacramento, California, February 1994).

17. California Air Resources Board, "Zero-Emissions Vehicle Update" (technical support document, Sacramento, April 1994).

18. These findings, based on Mark Delucchi's analysis of greenhouse gas emissions (see Table 2-1), are widely accepted by industry and governments around the globe (see *Choosing an Alternative Fuel: Air Pollution and Greenhouse Gas Impacts* [Paris: OECD, 1993]). He considered all fuel-related activities from extraction to combustion, including energy consumed in vehicle manufacture. He compared the emissions of different options by creating a composite carbon dioxide–equivalent measure. The analysis covers all important greenhouse gases, including not only carbon dioxide but also nitrogen oxides, hydrocarbons, carbon monoxide, and chlorofluorocarbons. The numbers presented in Table 3-2 represent percentage changes in greenhouse gas emissions resulting from the substitution of an advanced (post-2000), battery-powered electric vehicle for a comparable gasoline vehicle.

19. M. DeLuchi, "Emissions of Greenhouse Gases from the Use of Transportation Fuels and Electricity" (Argonne National Laboratory, Center for Transportation Research, Argonne, Illinois, 1991, ANL/ESD/TM-22). For a summary of findings, see M. DeLuchi, "Greenhouse-Gas Emissions from the Use of New Fuels for Transportation and Electricity," *Transportation Research* 27A (3): 187–91 (1993).

20. CARB, "Electric Vehicle Technology and Emissions Update" (Sacramento, California, April 1992).

21. Dan McCosh, "We Drive the World's Best Electric Car," *Popular Science* 244 (1): 52–56 (January 1994).

22. Membership in battery consortia was initially limited by national origin. U.S. ABC, for example, allowed only U.S. companies to participate. In fall 1993, U.S. ABC relaxed its all-U.S. rule in awarding a $12 million contract to a German company for sodium-sulfur battery development, but with the proviso that the batteries be manufactured in the United States.

23. Based on a confidential industry study.

24. For example, see CARB, "Electric Vehicle Technology."

25. Cited in *Business Week* (May 30, 1994): 110.

26. Cited in William Diem, *Automotive News* (May 9, 1994): 45.

27. Cited in CARB, "Zero-Emissions Vehicle Update," p. 60.

28. The economic analysis of the future cost competitiveness of electric vehicles presented here is based on an exhaustive review of literature and many interviews with manufacturers by Dr. Mark Delucchi and colleagues at the University of California, Davis. Earlier versions of the results have been widely reviewed and published elsewhere. See Mark DeLuchi, "Hydrogen Fuel-Cell Vehicles" (report 92-14, University of California, Davis, Institute of Transportation Studies, September 1992). See also Mark DeLuchi and Joan M. Ogden, "Solar Hydrogen Transportation Fuels," *Transportation Research* 27A: 255–75 (1993); International Energy Agency, *Electric Vehicles: Technology, Performance, and Potential* (Paris: OECD, 1993); and Q. Wang and M. A. DeLuchi, "Impacts of Electric Vehicles on Primary Energy Consumption and Petroleum Displacement," *Energy* 17: 351–66 (1992).

29. Other studies come to diverse conclusions. CARB, "Zero-Emissions Vehicle Update," projects even lower costs for electric vehicles, and a U.S. Department of Energy (DOE) study arrives at similar findings. Studies commissioned by major automakers are less optimistic, but they do not provide detailed cost breakdowns. See DOE, "Encouraging the Purchase and Use of Electric Motor Vehicles" (draft report, prepared by Abacus Technology Corporation, Washington, D.C., June 1994). See also Sierra Research and Charles River Associates, "The Cost-Effectiveness of Further Regulating Mobile Source Emissions."

30. Andrew Ford, "Electric Vehicles and the Electric Company," *Energy Policy* (forthcoming). Some details in the text are drawn from a longer unpublished report available from A. Ford at Washington State University.

31. By not increasing capacity I don't mean not building new capacity. Many utilities will find it to their advantage to build newer, cleaner, and more efficient power plants that are better suited to the smoother demand patterns of a system supplying electric vehicles, and to retire older, dirtier, and less efficient plants.

32. D. R. Buist, "An Automotive Manufacturer's Alternative Fuel Perspective," *Proceedings of the First Annual World Car 2001 Conference* (1993): 51–55.

33. D. Bunch, M. Bradley, T. Golob, *et al.*, "Demand for Clean-Fuel Vehicles in California: A Discrete-Choice Stated Preference Pilot Project," *Transportation Research* 27A (3): 237–54 (1993); J. E. Calfee, "Estimating the Demand for

Electric Automobiles Using Fully Disaggregated Probabilistic Choice Analysis," *Transportation Research* 27A (3): 237–53 (1985).

34. Ibid.

35. T. Turrentine, D. Sperling, and K. Kurani, "Market Potential of Electric and Natural Gas Vehicles" (report RR-92-8, University of California, Davis, Institute of Transportation Studies, 1992).

36. K. Kurani, T. Turrentine, and D. Sperling, "Demand for Electric Vehicles in Hybrid Households," *Transport Policy* (forthcoming).

37. K. Kurani, T. Turrentine, and D. Sperling, "The Zero Emissions Vehicle Market: A Diary-Based Survey of New Car Buyers in California" (Institute of Transportation Studies, University of California, Davis, preliminary report prepared for CARB, May 1994).

38. James P. Womack, "The Real EV Challenge: Reinventing an Industry," *Transport Policy* (forthcoming).

Chapter 4

1. EPA report 420-R-93-007, 1993, cited in Elmer W. Johnson, *Avoiding the Collision of Cities and Cars: Urban Transportation Policy for the Twenty-first Century* (Chicago: American Academy of Arts and Sciences, 1993).

2. U.S. Federal Highway Administration, "Summary of Trends" (Washington, D.C., 1990), p. 14.

3. William L. Garrison and J. Fred Clarke, "Studies of the Neighborhood Car Concept" (report 78-4, University of California, Berkeley, College of Engineering, 1977).

4. Michael Quanlu Wang, personal communication with author, August 1993. The information in the text is based on runs of models documented in Q. Wang, M. DeLuchi, and D. Sperling, "Emission Impacts of Electric Vehicles," *Journal of the Air and Waste Management Association* 40: 1275–84 (1990).

5. Automated vehicle control for conventional cars is already a primary focus of research being done at the University of California PATH program and by the Intelligent Vehicle Highway Systems program of the U.S. Department of Transportation, as well as by many companies. In theory, because computers can steer more precisely and brake more quickly than humans, a vehicle with automated controls would allow narrower lanes and more tightly packed cars. In practice, it might be preferable to have partially automated controls, with the driver capable of engaging them, for example by gripping the steering wheel.

6. C. Bosselman, D. Cullinane, W. Garrison, *et al.*, "Small Cars in Neighborhoods" (report PRR-93-2, Institute of Transportation Studies, University of California, Berkeley, 1993); A. Stein, K. Kurani, and D. Sperling, "Roadway Infrastructure for Neighborhood Electric Vehicles," *Transportation Research Record* (forthcoming).

7. Lloyd W. Bookout, "Neotraditional Town Planning," *Urban Land* 51 (1): 20–25 (January 1992); Peter Calthorpe, *The Next American Metropolis* (New

York: Princeton Architectural Press, 1993); Sim Van der Ryn and Peter Calthorpe, *Sustainable Communities* (San Francisco: Sierra Club, 1986); Jay Wallijasper, "At Home in an Eco-Village," *Utne Reader* (May–June 1992): 142–43; Wolfgang Zuckerman, *End of the Road: From World Crisis to Sustainable Transportation* (Post Mills, Vermont: Chelsea Green, 1991).

8. Energy consumed to heat and cool the interior of a small electric vehicle can be reduced with innovative techniques not suitable for larger vehicles on longer trips. Examples are solar-powered circulation and dry ice for cooling, and heated seats for warming.

9. From 1998 to 2002, the mandate will affect only automakers with annual sales of 35,000 in California. In 2003, the threshold will drop to sales of 3,000, encompassing many more manufacturers. Other states will probably use the same thresholds.

10. F. T. Sparrow and R. K. Whitford, "The Coming Mini/Micro Car Crisis: Do We Need a New Definition?" *Transportation Research* 18A: 289–303 (1984); T. Lipman, K. Kurani, and D. Sperling, "Regulatory Impediments to Small Electric Vehicles," *Transportation Research Record* (forthcoming).

11. Robert Wrede, "Sub-Car Electric Vehicle Opportunity" (report on workshop held at AeroVironment, Monrovia, California, 2 June 1993), appendix D.

12. K. Kurani and T. Turrentine, "Market Demand for NEVs" (report forthcoming from Institute of Transportation Studies, University of California, Davis).

13. A. B. Lovins, J. W. Barnett, and L. H. Lovins, "Supercars: The Coming Light-Vehicle Revolution," *Proceedings of the European Council for an Energy-Efficient Economy* (Rungstedgard, Denmark, June 1–5, 1993).

14. Seiss (1991), cited in ibid.

15. Amory B. Lovins, "Reinventing Wheels," *Atlantic Monthly* (forthcoming).

Chapter 5

1. See A. J. Appleby and F. R. Foulkes, *Fuel Cell Handbook* (New York: Van Nostrand Rheinhold, 1989); "Progress in Fuel Cell Commercialization," *Journal of Power Sources: Proceedings of the Second Grove Fuel Cell Symposium* 37 (1992); Mark A. DeLuchi, "Hydrogen Fuel-Cell Vehicles" (report 92-14, University of California, Davis, Institute of Transportation Studies, September 1992); David H. Swan and A. J. Appleby, "Fuel Cells for Electric Vehicles: Knowledge Gaps," *Proceedings of the Urban Electric Vehicle Conference* (Paris: OECD, 1992).

2. Fuel cells will also be used for electricity generation in stationary power plants. Fuel cells, in addition to being quiet and nonpolluting, have the advantage of being suitable for modular and small units. They can be part of a decentralized electricity-generation system, obviating the need for large centralized plants and expensive transmission lines. Several demonstration plants already exist. In early 1994 IFC, a unit of United Technologies, was installing fifty-eight 200-kw phosphoric-acid fuel cells worldwide.

3. D. H. Swan, B. E. Dickenson, and M. P. Arikaro, "Proton Exchange Membrane Fuel Cell Characterization for Electric Vehicle Applications" [SAE (Society of Automotive Engineers) paper 940296, Warrendale, Pennsylvania, 1994].

4. A. J. Appleby and F. R. Foulkes, *Fuel Cell Handbook* (New York: Van Nostrand Rheinhold, 1989), p. 190.

5. See Keith Prater, "Solid Polymer Developments at Ballard," *Journal of Power Sources* 37: 181–88 (1992).

6. Joan Ogden and Robert H. Williams, *Solar Hydrogen: Moving Beyond Fossil Fuels* (Washington, D.C.: World Resources Institute, 1989); and Joan Ogden, Eric Larson, and Mark Delucchi, "A Technical and Economic Assessment of Renewable Transportation Fuels and Technologies" (report prepared for Office of Technology Assessment, U.S. Congress, 1994, available from Institute of Transportation Studies, University of California, Davis).

7. Ogden, Larson, and Delucchi, "A Technical and Economic Assessment." Note that MacKenzie cites much higher costs, drawing on current technology from an earlier report by Ogden (James J. MacKenzie, *The Keys to the Car: Electric and Hydrogen Vehicles for the 21st Century* [Washington, D.C.: World Resources Institute, 1994]. Here costs, based on *future* technology, are much lower.

8. Mark A. DeLuchi and Joan M. Ogden, "Solar-Hydrogen Fuel-Cell Vehicles," *Transportation Research* 27A (3): 255–75 (1993).

9. Ogden, Larson and Delucchi, "A Technical and Economic Assessment."

10. J. M. Ogden and J. Nitsch, "Solar Hydrogen," in T. B. Johannson *et al.*, eds., *Renewable Energy: Sources for Fuels and Electricity* (Washington, D.C.: Island Press, 1993), pp. 925–1009.

11. Daniel Sperling, *New Transportation Fuels: A Strategic Approach to Technological Change* (Berkeley and London: University of California Press, 1988).

12. Mark A. DeLuchi, "Hydrogen Vehicles: An Evaluation of Fuel Storage, Performance, Safety, Environmental Impacts, and Cost," *International Journal of Hydrogen Energy* 14 (2): 81–130 (1989).

13. DeLuchi, "Hydrogen Fuel-Cell Vehicles." MacKenzie, *Keys to the Car,* pp. 74–77.

14. MacKenzie, *Keys to the Car,* p. 69.

15. General Motors, Allison Gas Turbine Division, "Research and Development of PEM Fuel-Cell System for Transportation Applications" (initial conceptual design report, prepared for Office of Transportation Technologies, U.S. Department of Energy, Indianapolis, Indiana, November 1993).

16. Robert Williams, "Fuel Cells, Their Fuels, and the U.S. Automobile," *Proceedings of the First World Car 2001 Conference* (1993).

17. Sperling, *New Transportation Fuels*. U.S. Department of Energy, Office of Policy, Planning, and Analysis, "Assessment of Costs and Benefits of Flexible and Alternative Fuel Use in the U.S. Transportation Sector, Technical Report 3: Methanol Production and Transportation Costs" (DOE/PE-0093, Washington, D.C., 1989).

Chapter 6

1. For example, see A. F. Burke, "The Hybrid Test Vehicle (HTV): From Concept through Fabrication and Marketing" (General Electric report 82CRD174, June 1982); G. G. Harding, *et al.*, "The Lucas Hybrid Electric Car" [SAE (Society of Automotive Engineers) paper 830113, Warrendale, Pennsylvania, 1983]; J. S. Reuyl, "XA-100 Hybrid Electric Vehicle" (SAE paper 920440, Warrendale, Pennsylvania, 1992); A. S. Keller and G. D. Whitehead, "Performance Testing of the Extended-Range (Hybrid) G-VAN" (SAE paper 920439, Warrendale, Pennsylvania, 1992). Also in 1994, Chrysler was building a hybrid race car using a flywheel and natural gas–fueled gas-turbine engine.

2. A. F. Burke, "Electric Vehicle Propulsion and Battery Technology, 1975–95," *Proceedings of the 25th Intersociety Engineering Conference on Energy Conversion* (Reno, Nevada, 1990); H. Huang, M. Debruzzi, and T. Riso, "A Novel Stator Construction for High Power Density and High Efficiency Permanent Magnet Brushless DC Motors" (SAE paper 931008, Warrendale, Pennsylvania, 1993); G. H. Cole, "Comparison of the Unique Mobility and DOE-Developed AC Electric Drive Systems" (EG&G report DOE/ID-10421, Idaho Falls, Idaho, 1993).

3. A. Lovins, J. W. Barnett, and L. H. Lovins, "Supercars: The Coming Light-Vehicle Revolution," *Proceedings of the European Council for an Energy-Efficient Economy* (Rungstedgard, Denmark, 1993).

4. *Electronic Engine Controls: Design, Development, Performance* (SAE publication SP-848, Warrendale, Pennsylvania, 1991).

5. A. F. Burke, "Energy Storage Specification Requirements for Hybrid-Electric Vehicles" (EG&G report EGG-EP-10949, Idaho Falls, Idaho, 1993).

6. R. F. Post, T. K. Fowler, and S. F. Post, "A High Efficiency Electrochemical Battery" (report UCRL-JC-110861, Lawrence Livermore National Laboratory, Livermore, California, June 1992).

7. A. F. Burke. "The Development of Ultracapacitors for Electric and Hybrid Vehicles: The DOE Program and the Status of the Technology" (SAE publication P265, Warrendale, Pennsylvania, 1993); A. P. Trippe, *et al.*, "Improved Electric Vehicle Performance with Pulse Power Capacitors" (SAE paper 931010, Warrendale, Pennsylvania, 1993).

8. D. C. LeBlanc, M. Meyer, Michael Saunders, *et al.*, "Carbon Monoxide Emissions from Road Driving: Evidence of Emissions Due to Power Enrichment," *Transportation Research Record* (forthcoming).

9. R. Mackay, "Hybrid Vehicle Gas Turbines" (SAE paper 93044, 1993); R. E. Symth, "Advanced Turbine Technology Applications Project Progress in Five Years" (SAE publication P265, Warrendale, Pennsylvania, 1993).

10. A. F. Burke, "Development of Test Procedures for Hybrid-Electric Vehicles" (EG&G report DOE/ID-10385, Idaho Falls, Idaho, 1992).

11. See G. H. Cole, "SIMPLEV: A Simple Electric Vehicle Simulation Program,

Version 1" (EG&G Idaho Report DOE/ID-10293, Idaho Falls, Idaho, 1991), and Version 2 (DOE/ID-10293-2, 1993).

12. The total energy used was calculated as follows:

$(kwh/km)_{total\ at\ source}$ = (EF) $(kwh/km)_{bat}$ + $26.8/mpg_{hybrid\ mode}$.

In this equation, 1 gallon of gasoline = 36.7 kwh and 15 percent of crude oil energy is lost in processing and distribution, where electricity generation and distribution efficiencies are as follows: 0.92 for distribution; 0.33 for power plants in 1992 and 0.45 in the future; 0.90 for battery chargers in 1992 and 0.92 in the future; and 0.80 for battery efficiency in 1992 and 0.85 in the future.

EF = $1/(eff_{distrib} \times eff_{powerplant} \times eff_{charger} \times eff_{battery})$.

EF = 4.55 in 1992 and 3.12 in the future.

13. A. F. Burke and K. Heitner, "Test Procedures for Hybrid-Electric Vehicles Using Different Control Strategies," *Proceedings of the 11th International Electric Vehicle Symposium* (Florence, Italy, 1992).

14. Lovins, Barnett, and Lovins, "Supercars."

15. Emissions of battery-powered electric vehicles were compared to gasoline vehicles in Chapter 3. It is important to recall that the estimated emissions are region specific, that is, highly sensitive to the type of power plant providing electricity and, if the plant is burning fossil fuels, to the stringency of emission controls on the power plant. For instance, as noted in Chapter 3, most electricity used in southern California comes from zero-emission, out-of-basin or well-controlled natural-gas power plants; thus electric-vehicle emissions there would be very low—much lower than reported in column 1 of Table 6-2—underscoring the variability of emission effects across regions.

16. Burke and Heitner, "Test Procedures"; A. F. Burke, "Development of Test Procedures for Hybrid-Electric Vehicles"; A. F. Burke. "On-Off Engine Operation for Hybrid-Electric Vehicles" (SAE paper 930042, Warrendale, Pennsylvania, 1993).

17. National Research Council, *Rethinking the Ozone Problem in Urban and Regional Air Pollution* (Washington, D.C.: National Academy Press, 1991).

Chapter 7

1. Robert Brown, "National Conference of States on Building Code Standards" (presentation to the Electric Transportation Coalition, Washington, D.C., November 8, 1993).

2. For example, see Brian Cook, *Bureaucratic Politics and Regulatory Reform: The EPA and Emissions Trading* (New York: Greenwood Press, 1988).

3. Toni Harrington, Honda North America, personal communication, July 1992, cited in John DeCicco and Deborah Gordon, "Steering with Prices: Fuel and Vehicle Taxation as Market Incentives for Higher Fuel Economy," in Daniel Sperling and Susan Shaheen, eds., *Transportation and Energy: Strategies for a Sustainable Transportation System* (Berkeley, California: American Council for an Energy Efficient Economy, forthcoming).

4. Jack Keebler, "U.S. Shelves Question-Ridden Two-Strokes," *Automotive News* (July 4, 1994): 10–11.

5. David L. Greene, "CAFE or Price: An Analysis of the Federal Fuel Economy Regulations and Gasoline Prices on New Car MPG, 1978–89," *Energy Journal* 11 (3): 37–57 (1990).

6. Oak Ridge National Laboratory, *Transportation Energy Data Book,* 13th ed. (Springfield, Virginia: NTIS, 1993).

7. In some areas, city government provides the electricity itself, with elected or appointed boards overseeing electricity supply. These municipal utilities are more easily encouraged to pursue broader social goals than are regulated companies. Two of the five largest municipal utilities in the United States are in Los Angeles and Sacramento, and both have been particularly aggressive at promoting electric vehicles.

8. California Public Utility Commission, "Opinion: Low Emission Vehicle Policy Guideline" (decision 93-07-054, San Francisco, July 21, 1993), p. 34.

9. Daniel Duann and Youssef Hegazy, *Natural Gas Vehicles and the Role of State Public Service Commissions* (Columbus, Ohio: The National Regulatory Research Institute, 1992), p. ix.

10. CARB regulators, however, severely restricted banking by confiscating a large portion of banked emissions after just one year. This is unfortunate because banking adds flexibility to the regulations, can reduce the cost of compliance by as much as 4 percent, and most importantly, is beneficial to the environment. By encouraging early cleanup of emissions, banking promotes cleaner air today when regulations are less stringent and harm per unit is greatest, and allows greater emissions in the future when standards will be more stringent and per-unit damage lower. See Jonathan Rubin and Catherine Kling, "An Emission Saved Is an Emission Earned: An Empirical Study of Emission Banking for Light-Duty Vehicle Manufacturers," *Journal of Environmental Economics and Management* 25 (3): 257–74 (1993).

11. The one other vehicle-emission averaging, banking, and trading program is an EPA program adopted in July 1990 for heavy-duty-engine manufacturers. It affects a relatively small number of vehicles. The effectiveness of this program has not been evaluated.

12. Michael Quanlu Wang, "Cost Savings of Using a Marketable Permit System for Regulating Light-Duty-Vehicle Emissions," *Transport Policy* (forthcoming).

13. Q. Wang, D. Sperling, and C. Kling, "Light-Duty-Vehicle Exhaust Emission Control Cost Estimates Using a Part-Pricing Approach," *Journal of the Air and Waste Management Association* 143: 1461–71 (1993).

14. W. Davis, M. Levine, and K. Train, "Feebates: Estimated Impacts on Vehicle Fuel Economy, Fuel Consumption, CO_2 Emissions, and Consumer Surplus (draft report, Lawrence Berkeley Laboratory, Berkeley, California, 1993).

15. Daniel Fessler, "Incentives for Improved Asset Utilization: The Regulator's Perspective" (paper presented at the RETSIE Conference, 11 June 1993), p. 7.

Chapter 8

1. See, for example James Fallows, *Looking at the Sun: The Rise of the New East Asian Economic and Political System* (New York: Pantheon Books, 1994).

2. This case study was based on a series of interviews conducted in 1993 and early 1994. Several sources preferred to remain anonymous and therefore names are withheld here.

3. This assessment of the views of automakers is based on hundreds of media interviews and personal conversations with executives and analysts from European, Japanese, and American auto companies, and related industries and government.

4. Sierra Research and Charles River Associates, "The Cost-Effectiveness of Further Regulating Mobile Source Emissions" (report prepared for the American Automobile Manufacturers Association, Sacramento, California, 1994); and "Economic Consequences of Adopting California Programs for Alternative Fuels and Vehicles" (report by DRI/McGraw-Hill and Charles River Associates, Washington, D.C., 1994). See also Dean Drake, "Technology, Economics, and the ZEV Mandate," in D. Sperling and S. Shaheen, eds., *Transportation and Energy: Strategies for a Sustainable Transportation System* (Washington, D.C. and Berkeley, California: American Council for an Energy-Efficient Economy, forthcoming).

5. See, for example, Douglas Lavin, "Robert Eaton Thinks Vision Is Overrated and He's Not Alone," *Wall Street Journal,* October 4, 1993.

6. M. Granger Morgan and Robert M. White, "A Design for New National Laboratories," *Issues in Science and Technology* (Winter 1993–94): 29–32.

7. U.S. Congress, Office of Technology Assessment, *Defense Conversion: Redirecting R&D* (Washington, D.C.: Government Printing Office, May 1993).

8. Ibid., p. 15.

9. This case study is based on a series of interviews with the principal developers (William Falik, Tom Reiman, and Jonathan Cohen), the staff of Sutter Bay Associates, and a series of accounts in local newspapers.

Index